T0142904

Cyber Security Meets Machine Learning

Xiaofeng Chen • Willy Susilo • Elisa Bertino
Editors

Cyber Security Meets Machine Learning

Springer

Editors
Xiaofeng Chen
School of Cyber Engineering
Xidian University
Xi'an, Shaanxi, China

Willy Susilo
School of Computing and Information
Technology
University of Wollongong
Wollongong, NSW, Australia

Elisa Bertino
Department of Computer Science
Purdue University
West Lafayette, IN, USA

ISBN 978-981-33-6728-9   ISBN 978-981-33-6726-5   (eBook)
https://doi.org/10.1007/978-981-33-6726-5

This Springer imprint is published by the registered company Springer Nature Singapore Pte Ltd.
The registered company address is: 152 Beach Road, #21-01/04 Gateway East, Singapore 189721,
Singapore

Preface

Cyber threats are growing in complexity. Hundreds of millions of new strains of malware are identified each year. New types of malware programs can avoid detection by traditional anti-virus or even operate without using binary files at all (i.e., fileless attacks). Attacks are becoming more multilayered, involving a combination of network-based techniques, malware, and web application attacks. Insider threats are a growing problem, and insider attacks are very difficult to distinguish from legitimate user activity. Attackers are also leveraging devices such as mobile phones, connected devices in the office and home, and IoT infrastructure to carry out large-scale attacks. Machine learning algorithms can help detect and mitigate many of these new threats. They are able to analyze a much larger volume of data than human security professionals, intelligently identify anomalies and suspicious behavior, and investigate threats by correlating many data points. However, the results given by the machine learning model are not entirely credible. Furthermore, the vulnerability of the machine learning models against adversarial attacks is a major issue of artificial intelligence technologies, not to mention that the privacy of the data used in the training and testing is a major users' concern as well. The current book, *Cyber Security Meets Machine Learning*, timely focuses on those critical issues by covering research advances on the following areas: Cyber Security based on Machine Learning, Security in Machine Learning Methods and Systems, and Security and Privacy in Outsourced Machine Learning.

The book comprises 6 chapters, written by 22 authors who are active researchers or practical experts in areas related to cyber security and machine learning technologies. Although the authors are from different areas and subareas, they share a common goal: design effective approaches to secure the cyberspace.

Chapter "IoT Attacks and Malware," by Anand Mudgerikar and Elisa Bertino, opens with a survey of different types of attacks on various IoT devices according to the goals of the attackers, including Passive/Information Stealing Attacks, Service Degradation Attacks, and Botnet-Based Attacks, followed by a discussion on machine learning-based security solutions.

Chapter "Machine Learning-Based Online Source Identification for Image Forensics," by Yonggang Huang, Lei Pan, Wei Luo, Yahui Han, and Jun Zhang, presents a novel scheme based on machine learning for online identification of image sources, especially images shot by unknown camera models.

Chapter "Reinforcement Learning Based Communication Security for Unmanned Aerial Vehicles," by Liang Xiao, Donghua Jiang, and Sicong Liu, introduces a study of a reinforcement learning-based scheme for UAV transmission against jamming attacks, which improves the secrecy capacity of the UAV system against smart attackers.

Chapter "Visual Analysis of Adversarial Examples in Machine Learning: A State-of-the-Art and Future Trend," by Wei Zong, Yang-Wai Chow, and Willy Susilo, presents an overview of research on adversarial examples, studying the tools for visualizing the generation of adversarial examples and the methods of detecting adversarial examples while investigating the means for improving the robustness of the machine learning model.

Chapter "Adversarial Attacks Against Deep Learning-Based Speech Recognition Systems", by Xuejing Yuan, Yuxuan Chen, Kai Chen, Shengzhi Zhang, and XiaoFeng Wang, details an approach to attack Automatic Speech Recognition systems in the real world by utilizing generated adversarial audio examples, which succeeded against Google Assistant, Google Home, Amazon Echo, and Microsoft Cortana.

Chapter "A Survey on Secure Outsourced Deep Learning," by Xu Ma, Xiaoyu Zhang, Changyu Dong, and Xiaofeng Chen, presents a comprehensive view on outsourced computation in deep learning, analyzing the underlying cryptography techniques and outsourced architectures concerning efficiency, security, and privacy and providing the insights of research issues to be addressed in the future.

Overall, this book makes a solid contribution to cyber security and machine learning area, not only helping develop suitable machine learning tools effective against cyber threats but also inspiring researchers to carry out analyses and experiments to identify vulnerabilities in existing machine learning techniques and improve the adversarial robustness of models. The editors are confident that this book will significantly contribute toward the challenging field of cyber security and machine learning.

We would like to conclude this preface with our acknowledgments. First and foremost, we would like to thank the contributors to this book for their support and patience. We are also very grateful to the team from Springer for their dedication in putting together this significant book. This book is supported by the National Natural Science Foundation of China (no. 61960206014) and China 111 Project (no. B16037).

Xi'an, China   Xiaofeng Chen
Wollongong, NSW, Australia   Willy Susilo
West Lafayette, IN, USA   Elisa Bertino
November 5, 2020

Contents

IoT Attacks and Malware

Anand Mudgerikar and Elisa Bertino

1 Introduction

As of today, there are around 5.8 billion IoT devices or end points. This number
is estimated to grow to 30 billion in 2021. These devices are part of the "Internet
of Things" (IoT), which is defined as a "system of interrelated computing devices,
mechanical and digital machines, objects, animals or people that are provided with
unique identifiers (UIDs) and the ability to transfer data over a network without
requiring human-to-human or human-to-computer interaction [55]." Along with
IoT, advances in other technologies like cloud/edge computing [49], embedded
devices [34], and machine learning [36] have enabled the creation of complex,
intelligent, and autonomous IoT ecosystems. Such IoT ecosystems consist of
multiple devices interacting with one another using various services, APIs, apps,
etc. All components of such an ecosystem typically work asynchronously in order
to run efficiently. The basic methodology toward executing a command, such as
turning a light switch on, is "fire and forget." Therefore, each IoT device takes
actions to achieve local goals but without considering the global environment, which
leads to safety and security issues. In addition, many IoT devices are designed
with poor or no security mechanisms in place. Along with this, the huge number
of devices and heterogeneity in terms of functions, protocols, manufacturers, etc.
add to the complexity. All these factors lead to security, safety, and privacy issues in
IoT ecosystems.

Due to the heterogeneity and complex nature of IoT systems, traditional security
approaches seem infeasible. A possible solution is to design artificial intelligence
(AI)-based security solutions able to handle complexity. Notable examples of AI-
based security solutions are the intrusion detection/prevention systems (IDSs/IPSs)

A. Mudgerikar · E. Bertino (✉)
Purdue University, West Lafayette, IN, USA
e-mail: amudgeri@purdue.edu; bertino@purdue.edu

that are able to detect patterns in network traffic, system behavior, access control, service usage, etc. [30, 66] either to match with attack signatures or to report the behavior as anomalous. There have been attempts to build such AI-based IDS for IoT networks [15]. Most of these intelligent IDSs build patterns or models for steps of the attack rather than the entire attack which would be infeasible, essentially breaking down an attack into steps or "kill chain." So, in order to build better AI-based attack models for IoT networks, we need to classify and analyze IoT attacks as kill chains rather than attack instances.

A kill chain makes detection feasible by breaking down an attack into steps, so that the AI-based IDS can detect patterns to match these steps rather than the entire attack. The general approach to classify and structure attacks in traditional networks is through a cybersecurity kill chain like the ones by Lockheed Martin [43], Gartner [12], and MITRE [58]. Those kill chains break down the attack into the following steps: reconnaissance, intrusion, exploitation, lateral movement, obfuscation, and finally ex-filtration. Some of the chains combine some steps into one or break one into several steps.

In traditional security solutions, the focus for detecting attacks is on the first 3 steps of the attack kill chain, namely reconnaissance, intrusion, and exploitation. This is a reasonable choice as the attacks should be detected as soon as possible and detection becomes more difficult as the attacker "gains a better foothold" in the system through rootkits and obfuscation techniques. However, in an IoT environment, detection in these initial steps is very difficult because of two factors: heterogeneity and vulnerability of devices. Because of those factors, signature datasets of exploits are huge in size, which results in large numbers of false positives. Another issue is that, since so many IoT devices have weak passwords or other well-known security issues, it is difficult to differentiate between benign usage and attacks in just these three steps. Also, many IoT-based attacks like distributed denial of service (DDoS) generally do not follow the trend of traditional kill chains as they directly move from step 1 to step 6. This is the case, for example, IoT used as bots in a botnet to launch attacks.

IoT devices have been exploited in several large-scale DDoS attacks, like the attack on Dyn DNS and other major websites which reached a peak of 1.1 Tbps (terabits per second) and involved 1.48 million compromised IoT devices. Botnets are not a new problem, and there have been numerous instances of botnets in the past. However, with the exponential growth of IoT devices with weak security mechanisms, these botnets have become very powerful and can be used to perform massive-scale DDoS attacks. In February 2018, there was the largest scale DDoS attack, known as "MemCrashed" [40], which reached a peak of 1.35 Tbps using amplification techniques and exploiting the "Memcached" service. DDoS attacks are the most popular form of attacks that exploit IoT devices, but these devices are vulnerable to other forms of attacks. Many IoT devices include smart home devices (home appliances like fridges, TV, cameras, etc.), personal/wearable devices (like wristwatches, fitness devices, music players, etc.), and medical devices used in hospitals (heart rate monitors, implants, skin sensors, etc.). Such kind of devices operates on and stores private information and is thus prone to privacy breaching attacks. However, IoT malware has since evolved to not only steal the data but also lock devices and demand ransoms. Unlike traditional ransomware, IoT ransomware, also called "Jack-ware," performs full-disk encryption which intends to completely lock the device until the ransom is paid rather than encrypting particular files on the system. Recently, a number of IoT devices, like thermostats, were shown to be vulnerable to such ransomware. Such jack-ware could lock IoT devices like cars, TVs, thermostats, etc. and could prove to be even more harmful than traditional ransomware.

IoT devices with embedded sensors are widely used in critical infrastructure systems; for example, smart meters used to monitor and manage power consumption in buildings. Attackers have exploited such vulnerable smart meters to underreport the power usage [65]. Similarly, smart homes devices, like home heating systems (increase heating or turn it off) and smart locks (gaining unauthorized entry to homes), have been exploited to cause damage [23]. Another example is attackers compromising a smart vehicle and manipulating the braking/steering of the vehicle [50]. Wireless sensor networks are also been slowly integrated into the IoT environment with protocols such as 6LowPan and CoAP. Such wireless sensors with poor security mechanisms provide the perfect opportunity for attackers to disrupt industrial and enterprise systems. A recent alert issued by FDA reported vulnerabilities in Abott's implantable cardiac pacemakers that allowed attackers to send malicious commands to the device [14].

There have been numerous instances of service degradation attacks on wireless sensor networks (WSNs). More recently, a massive attack, called "brickerbot" [9], infected 2 million IoT devices. It completely wiped the firmware on these devices and replaced it with random data.

The previous discussion clearly shows that the attackers have different motivations and goals for compromising IoT devices. We see that the intrusion points for all these attacks are often the same, such as known vulnerabilities in devices, weak passwords, etc. However, the goals of the attackers are often different. This means that the first three steps of the kill chain are similar for most of those attacks, but the last three steps are highly different. Therefore, the goal of this chapter is to take an important initial step toward the design of scalable AI solution for IoT ecosystems by classifying IoT attacks based on the last three steps of the kill chain, namely lateral movement, obfuscation, and ex-filtration, collectively referred to as "goals of the attacker."

In this chapter, we survey attacks on/using IoT devices and classify them into following categories according to the goals of the attacker:

- *Passive/Information Stealing Attacks:* The goal of the attacker is to steal important or private information stored or being communicated from IoT devices.
- *Service Degradation Attacks:* The goal of the attacker is to deny or degrade the services provided by IoT devices.

– *Botnet-Based Attacks:* The goal of the attacker is to take control of IoT devices and use them as bots in attacker-controlled botnets. They can be used to launch large-scale DDoS attacks or perform other malicious activities such as crypto-mining. The goal of our classification is to help build better attack profiles and models for AI-based IDSs and IPSs.

This chapter is organized as follows. In Sect. 2, we provide relevant background on traditional cybersecurity kill chains and major IoT security weaknesses. In Sect. 3, we discuss our classification scheme in detail and survey different attacks from each attack category. In Sect. 4, we perform an analysis of IoT malware in terms of system calls. In Sect. 5, we discuss related work and how AI-based security solutions can benefit from our classification work. We outline some conclusions and future work in Sect. 6.

2 Background

In this section, we discuss the major steps involved in attacks and give some examples of popular cybersecurity kill chains. After which we analyze the major security weaknesses in IoT-based networks due to which detection is hard in the initial steps of the kill chain.

2.1 Cybersecurity Kill Chains

There are a number of popular kill chains with different number of steps, but essentially all these chains can be considered as organized into the six following steps:

1. *Reconnaissance:* In this step, the attacker gathers information about the network and environment which will later help in the attack. This step generally involves detecting potential victim devices, network sniffing, port scanning, etc.
2. *Intrusion:* In this step, the attacker gains access to the network or device using some vulnerabilities like weak passwords, known system vulnerabilities, malware, etc. This is the stage by which the attacker establishes a channel between himself/herself and the victim.
3. *Exploitation:* In this step, the attacker establishes a foothold in the network by exploiting the device. It typically involves installing some form of rootkits or malware through ftp or drive by downloads after the initial intrusion is successful. In IoT networks, this step would download a device's CPU architecture-specific malware binary on the target device. Now, the attacker has complete control of that particular device in the network.
4. *Lateral Movement:* In this step, the attacker attempts to infect more systems in the network and execute more fine-grained active reconnaissance as now the

attacker is posing as a legitimate device. This step usually involves spreading the malware to other vulnerable devices by probing, brute-forcing passwords, etc. It could also involve escalating privileges in the device itself, by, for example, identifying stored passwords or sensitive user data on the device itself. Again, now the attacker can pose as a legitimate device to gain application sensitive data on the network like certificates, keys, user data/credentials, etc.

5. *Obfuscation:* In this step, the attacker tries to hide from security services running on the network. This involves clearing logs, rootkits, uninstalling security services, etc.

6. *Ex-filtration:* In this step, which is the final one, the attacker proceeds to achieve the goal or main function of the attack. This could be denial of service (DOS), DDoS, stealing information, etc.

2.2 Major IoT Security Concerns

We now analyze some of the known design decisions of IoT networks which make it easier for the attacker to infect or gain access to devices. In fact, these are the main reasons for which detection in first three steps of a kill chain is difficult.

Exposure to the Internet One of the main reasons for the ease of large-scale exploitation of IoT devices is the exposure of IoT devices on the Internet, which makes it very easy to locate them. For example, Shodan is a search engine that allows the user to find different types of IoT devices accessible over the Internet. It returns data about standard services such as web servers (HTTP ports), FTP, SSH, Telnet, SNMP, IMAP, SIP, and RTSP. A Shodan scan executed in 2016 across the USA returned 1.78 million records with 0.45 million unique IPv4 addresses and 0.26 million IPv6 unique addresses. These devices are not behind a firewall as they should be, but rather accessible remotely over the internet. This is a major security concern. Another issue with IoT networks is that they expose services over the Internet which were designed for local networks and never meant to be directly accessible from Internet. For example, the "memcached" service is a local network caching service that allows database-driven websites to cache data and objects. Many servers misconfigure this service and allow attackers to access this service over the Internet. Such a misconfiguration allowed attackers to use this service to launch the massive 1.35 Tbps DDoS attack on GitHub [40].

Host Vulnerabilities Even though standard operating system and protocol vulnerabilities are being patched in IoT devices, there are still numerous vulnerabilities in the firmware and application layers of these devices. Another problem is that IoT devices rarely update their firmware and applications and continue to run insecure versions. This is evident in a recent vulnerability analysis of 1501 services in 12 different institutions using ShoVAT [19], in which 3922 known vulnerabilities were identified. These known vulnerabilities were from the list of common vulnerabilities in the national vulnerability database (NVD) maintaining the common vulnerabilities and exposures (CVEs). It is important to note that there are also numerous zero-day or unknown vulnerabilities that can also be used to attack the devices.

Weak Passwords One of the main injection points for IoT devices is the use of password-based offline dictionary attacks. Most IoT devices use default or weak passwords which make them easy targets for compromise even if no vulnerabilities are known or used by the attackers. The open SSH and telnet ports give the attackers the perfect opportunity to gain root access to the devices using password dictionaries. The most powerful DDoS malware Mirai [67] used a dictionary of 60 common/default factory passwords, and it was able to compromise more than 49,000 devices.

3   Attack Classification

We surveyed and analyzed different kinds of attacks on IoT devices that have been reported over the years. We found that all of the analyzed attacks can be classified into the three main categories defined before as shown in Fig. 1. In this section, we categorize and analyze the attacks based on the differences specifically in the final steps of the kill chain, that is, lateral movement, obfuscation, and ex-filtration rather than the initial infection steps.

3.1   Passive/Information Stealing Attacks

These kinds of attacks include passive sniffing and eavesdropping techniques. The goal of the attacker is to obtain private information. Such privacy breaching attacks are generally used to attack smart home devices and wearable health care/monitoring devices. Examples of offline information recorded are personal information like music tastes, sleep patterns, exercise routines, child behavior, medical information, etc. The most effective detection of such attacks is during the eavesdropping step, by which the attacker sniffs the data, or during the data transfer step, by which the attacker transfers data learnt from the network to some external server controlled by the attacker. Popular eavesdropping techniques include man-in-the-middle (MITM) attacks, network traffic sniffing, masquerading as fake gateways, forging, etc. Generally, attackers use the file transfer protocol (TCP), IRC channels, HTTP, etc. to transfer data. The attacker's goal is to masquerade its traffic as benign. Generally, in IoT environments, there is a regular transfer of benign data collected from devices to external servers using FTP or HTTP protocols. This allows the attacker to hide the malicious transfers among these benign transfers and make detection hard during the data transfer stage. So, it is more effective to detect such attacks during the eavesdropping step.

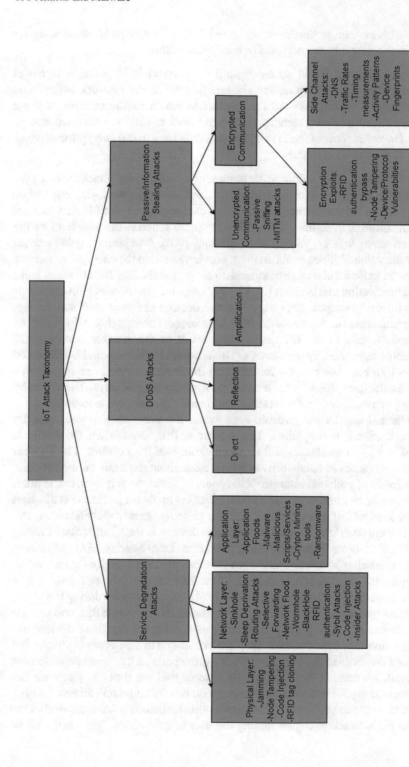

Fig. 1 IoT attack taxonomy

Those attacks can be further categorized into two categories depending on whether they are against unencrypted or encrypted traffic.

Unencrypted Traffic When no encryption mechanism is in place, it is trivial for the attacker to eavesdrop on the communication in the network using basic MITM attacks. These attacks are the hardest to detect as the attacker is using only passive eavesdropping techniques that are hard to differentiate from benign behavior. However, most of the IoT devices are switching to use encryption services like TLS to prevent such attacks.

Encrypted Traffic Encryption protects message contents by attackers that passively sniff the communication in the network. IoT networks deploying encryption are however still vulnerable to (1) encryption exploit attacks and (2) side-channel attacks. In addition, in some cases, weak encryption schemes are adopted (see the analysis on smart toys by Valente and Cardenas [60]). *Encryption exploit attacks* exploit some vulnerabilities to obtain the secret keys used in the encryption process. Some attacks exploit vulnerable communication protocols, like RFID, which lacks proper authentication mechanisms [5]. Device tampering techniques allow attackers to obtain the cryptographic keys stored on the devices and then sniff and decrypt the communication on the network. Any host-based vulnerability, which can be used to gain access to the IoT device, can result in the attacker obtaining the cryptographic keys used by the devices. The analysis by Valente and Cardenas on smart toys [60] has shown a case in which all the devices of the same type from a certain manufacturer shared a fixed small pool of hard-coded keys. Therefore, by just buying one such device, the attacker could easily gain access to these keys.

Side-channel attacks are possible even when the encryption system used by the network cannot be exploited. The reason is that, even when the traffic is encrypted, a lot of metadata is still exposed clearly on the network. The attacker can use this metadata to obtain private information about the users on the network. Common forms of such metadata are DNS requests, device fingerprints, traffic rates, request/response timings, etc., which allow attackers to obtain information like user behaviors, kind of IoT devices used, activity patterns, sensor classifications, etc. The most recent such attack on IoT smart home devices is the "Traffic Rate Privacy Attack" [3]. It involves identifying the devices from DNS queries, MAC addresses, and traffic rates after which user activities on these identified devices are recorded. For example, the attacker is able to infer when the user uses his/her microwave in the day or when the user listens to music on his/her Amazon echo during the week.

Intelligent virtual assistants (IVAs), like Amazon Alexa, Apple Siri, and Google home, have become very popular in recent years. Even though these devices use encrypted communication channels, they are vulnerable to side-channel attacks and vulnerable IVA-enabled devices. In addition, some parts of the communication are unencrypted. As these devices are usually listening all the time, they provide the perfect opportunity for attackers to eavesdrop on private communications. Another way of exploiting such devices is by impersonating malicious voice commands. One such example is attackers impersonating the user to gain access to a smart lock or other smart devices. These IVA devices do not have the capability to effectively differentiate between the user's voice commands and the attacker's impersonated voice commands.

3.2 Service Degradation Attacks

The goal of such attacks is to either completely stop the service from working properly (denial of service) or degrade the service performed by the device. These attacks can be further classified into three categories, based on network layer of the victim service or device, to which the attacker has access to, as follows: physical layer, network layer, and application layer attacks.

Physical Layer Attacks The prominent forms of such attacks are as follows:

Jamming In these attacks, the attacker interferes with the frequencies of the legitimate devices by introducing noisy traffic on these frequencies. Attacks have been shown by which the attacker sends noisy signals that cause interference on the RFID tags used by the sensors. The result is that the whole network becomes unusable as the legitimate devices cannot distinguish between the legitimate traffic and the injected noisy traffic.

Node Tampering In these attacks, the attacker is able to access and modify legitimate devices on the network. The attacker can modify the code on the device, insert malicious code onto the devices, or tamper with its circuitry. The compromised device then either stops functioning properly (DoS) or might be used to introduce noise in the network (jamming) which slows or brings the entire network down. This kind of jamming is even more difficult to detect as the source of the noise is a legitimate device.

RFID Tag Cloning RFID tags can be cloned by the attackers as most of the embedded devices lack proper authentication mechanisms. The attacker is able to inject a malicious clone into the network and, by using the clone, read, delete, or modify the data on the network. This way the attacker can manipulate the functionality of the network and cause the network to produce outputs as required by the attacker.

Code Injection In these attacks, the attacker injects malicious code on the device by physical means like plugging in an external device like a USB stick. This malicious code could be any form of malware like worms, viruses, ransomware, etc.

Network-Level Attacks In the network-level attacks, the attacker has access to the victim's network or is able to introduce malicious traffic in the network. The attacker does not have physical access to any of the devices but is able to inject and read traffic in the network. The prominent forms of such attacks are as follows:

Sleep Deprivation Attack   The attacker tries to make the legitimate devices use more power than usual by constantly sending requests to them. Most of the small IoT devices have sleep cycles to save power. The end result is that the device consumes all its power and shuts down causing a DoS.

Sinkhole Attack   The attacker manipulates the devices into sending all the packets into a sink device. Such a sink device is compromised by the attacker in most cases. The result is that communication in the network is disrupted as the intended receiver of the packets does not receive the packets.

Routing Attack   The attacker spoofs or modifies the routing information communicated in the network. This disrupts the network communication and can even cause routing loops.

Network Flooding Attack   The attacker sends a flood of request messages, like TCP SYN, HELLO, ACK, UDP, and ICMP messages, from a compromised device on the network or inserts messages with a spoofed source addresses in the network. The device being flooded runs out of memory while responding or allocating memory for these floods of requests.

Selective Forwarding Attack   The malicious devices on the network drop certain packets intentionally and not forward them to the base station. This disrupts the network communications as the base station is not receiving packets from some legitimate devices.

RFID Authentication Attacks RFID devices are usually resource-constrained devices and do not have proper authentication mechanisms in place. Attackers can then introduce malicious devices able to use this vulnerability to pose as legitimate devices on the network and thus can read, delete, and modify the communication in the network. There are numerous examples of attacks exploiting RFID authentication mechanisms and manipulating the network [37].

Sybil Attacks   The attacker forges identities of legitimate devices and introduces malicious devices with these forged identities. This is usually seen in peer-to-peer IoT networks. This allows the attacker to get a disproportionate control over the network and can allow the attacker to manipulate outcomes such as voting results, incorrect decisions, etc.

Code Injection   This is similar to code injection attack in the physical layer, but the attacker here exploits some network protocol vulnerabilities to inject malicious code into the devices like Heartbleed [20] against SSL/TLS, Beast attack against HTTPS [56], etc.

Insider Attack   In such attack, the attacker has already compromised a device in the network and then uses this device to perform attacks. Insider attacks have been a topic of ongoing research in the area of IoT security [51].

Wormhole Attack The attacker receives a packet at one end point of the network and then forwards or "tunnels" it to another point in the network using a powerful antenna or a colluding attacker. This results in all devices sending their packets through the attacker and not discovering any other legitimate but slower paths. The attacker thus gains complete control of the network communications and can disrupt communication by dropping all packets or dropping packets from certain devices.

Black Hole Attack It is also called packet drop attack. This kind of attack affects routers and forces the router to drop all the incoming packets instead of forwarding them.

Application-Level Attacks Application-level attacks focus on specific applications running on IoT devices. At the application level, sometimes, the goals of the attacker may not be purely denial of service but some other functions like crypto-mining, ransoms, etc. However, we categorize them here as they indirectly result in degradation of the services performed by the device by either consuming too many resources on the device like in crypto-mining or making the device unusable in the case of ransomware. The prominent forms of such attacks are as follows:

Application Flood Attack We discussed network-level flooding attacks previously. Similar flooding attacks are possible in the application layer, such as the Slowloris attack, HTTP layer 7 flood attacks, etc. These attacks focus on specific applications running on the IoT device and send a flood of requests to the application. These kinds of attacks are much harder to detect as all the requests look legitimate and firewalls cannot distinguish between legitimate requests and requests being sent by the attacker especially if the source IP address is spoofed.

Malicious Scripts and Malware The IoT device landscape is highly heterogeneous in terms of the device architectures, operating systems, applications, functionalities, and protocols. Therefore, many IoT devices have vulnerabilities that can be exploited by various malicious scripts, viruses, worms, spywares, etc. Various malicious scripts that exploit services like Active-X, JavaScript, etc. have been known to compromise IoT devices and manipulate their functions. These scripts are harder to detect than malware because they do not employ spreading mechanisms but just manipulate the data being sent or received by the IoT devices. Malware specifically tailored for IoT devices is also growing at a rapid pace, and there is little research in this field. The most recent examples being Reaper [31] and IoTroop [24]. These malware have recently been used for building botnets, which we discuss in detail in the next subsection. The reason for the lack of research can be attributed to the heterogeneous nature of IoT devices in terms of architectures, protocols, etc.; thus, each malware has numerous versions, and it is difficult to keep track of them. The trend in IoT malware is to stay dormant but keep spreading and compromising all possible devices on the network [32].

Crypto-Mining Malware Malware, such as the Darlloz Linux worm [38], are equipped with an open-source cryptocurrency mining tool called "cpuminer" and are being used for mining Bitcoins, Dogecoins, and Mincoins. This is a growing trend in IoT malware and can be seen as an addition to the latest versions of the popular IoT malware.

Ransomware This is the latest trend of malware that is emerging on IoT systems. Ransomware [11] or jack-ware can be used to lock the IoT devices used in hospitals, vehicles, etc. and demand ransoms to unlock them.

Bricking Attack As seen in the recent Brickerbot [9] malware, the attacker might aim to factory reset the device or render it completely useless by overwriting the operating system with random data. Brickerbot used Telnet, SSH, HTTP, HNAP, and SOAP as its attack vectors, which included some known vulnerabilities and simple password dictionary attacks [26]. The exploit code used to "brick" the device with random code is highly device specific, and the malware contains such code for a large number of device manufacturers including Ingenic devices, 3Com Access Points, Aver DVR, etc.

3.3 DDoS Attacks

DDoS attacks are the most powerful form of attack that exploit IoT devices with the latest attack reaching a peak of 1.35 Tbps [40]. The goal of the attacker is to grow and maintain a botnet of attacker-controlled devices which can be later used in launching massive-scale attacks. Since DDoS attacks have a unique kill chain compared to traditional attacks, we analyze these attacks according to the following three main stages:

1. **Injection Stage.** In this stage, the IoT devices are compromised typically using either some vulnerability or brute forcing the passwords. The goal of the attacker is to gain access to the device and install some malicious code on the device. The end result is that the attacker is able to convert the IoT device into a bot which is under the attacker's control. This stage is identical to the infection step of the previous attack categories and is similar in all IoT-based attacks.
2. **Communication Stage.** In this stage, the compromised bots communicate with the bot master or agents controlled by the bot master. This communication generally uses Internet Relay Chat (IRC) channels, custom web-based applications that use HTTPS encrypted communication, the XMPP open-source instant message protocol, P2P network communication like Bittorrent, etc. Some botnets have evolved to use standard social messaging websites, like Instagram, Facebook, etc., and TOR hidden services for communication and thus avoid being stopped by egress filtering on the network. The architecture of the communication system is generally an agent-handler relationship [2], where each bot is under the control of an agent and these agents are controlled by the bot master. The agent to bot communication may consist of various commands sent by the agent to the bot. These commands could be control messages to maintain the communication channel, attack messages to start an attack according to the parameters in the

message (start time, target IP, attack rate, etc.), network scan reports of the bot network which the attackers can use to find new vulnerable devices or potential bots on the network, etc. Some malware intentionally disable other services on the bot, so that the bot is not accessible by the user and other malware. This stage involves components of the lateral movement and ex-filtration steps of the kill chain.

3. **Attack Stage.** In this stage, a DDoS attack is performed by a number of bots. Each such bot will usually be performing a simple UDP, ICMP, SYN, ACK, and HTTP flood attack, or more complex application-level flood attacks. Different malware use different rates of attack. Some might use the maximum available bandwidth, while others are more conservative to avoid being detected. This stage involves the final ex-filtration step of the kill chain.

The final ex-filtration step, carried out in the attack stage, can be classified into the following three categories:

1. *Direct.* The bots directly send traffic to the victim. Most of the time, they spoof the source IP address to make it difficult to identify the bots. However, even with a spoofed IP address, it is relatively easy to find the bots by tracing the path taken by the packets. Due to these reasons, current attacks rarely use this strategy. All kinds of flood attacks, like TCP SYN, ACK, and Hello floods, are possible.

2. *Reflection.* The bots send spoofed requests to another host on the network. The spoofed address is the address of the victim. This results in all these hosts to send a flood of responses to the victim. Services that have been used for such attacks include NTP [45], SSDP [21], IP fragmentation, DNS, SRCDS [4], Chargen, Call of Duty, SNMP, CLDAP [8], Sunrpc [57], Netbios, HTTP, and RIP [41]. They are used to "reflect" these responses onto the victim. All these services are either misconfigured, vulnerable to exploits, or do not perform authentication properly, which makes it possible to exploit them for reflection/amplification purposes. Reflection-based attacks make it harder to find the compromised bots as the responses received by the victims are from the reflecting machines. This makes it difficult to detect the addresses of the bots especially if the reflector machines do not track or monitor the received requests. Most of the current DDoS attacks nowadays employ this strategy.

3. *Amplification.* This type of attack is very similar to reflection-based attack, but the main difference is in the bandwidth of the attack. In a reflection-based attack, the request to the response rate is one, meaning that there is generally one response of similar size from the reflector machine for one request. But in an amplification attack, the attacker tries to magnify the amount of traffic that is sent to the victim. A carefully crafted packet sent to the exploitable services used for reflection attacks can amplify a request bandwidth of 1 Gbps to 100 Gbps [40]. Popular examples of such attacks are DNS amplification attacks [1], which exploits vulnerable open DNS servers, and the recent "memcrashed," which exploits the "memcached" service open on Linux machines.

It is important to point out that those reflection and amplification attacks have used popular standard network services like DNS and SSDP. These attacks are easy to mitigate if the inbound traffic is filtered for these ports. But attackers have recently started switching to using application-level services, like memcached and Call of Duty servers [61], to perform these attacks. Application-level attacks are much difficult to filter as it is difficult to distinguish between legitimate traffic and malicious traffic. In the future, smart attackers will switch to more sophisticated attacks that would employ specially crafted application-level requests to carry out amplification and reflection attacks.

We now describe the main features of the most popular malware that has been used to launch IoT-based DDoS attacks. We categorize different malware into several families discussed in what follows. The main criterion for the classification is the different behavior in any of the three stages of infection: injection, communication, and attack. We give a summary of the different behavior of the families in Table 1.

Family 1 This family of malware started with the Linux.Hydra malware and evolved into a few variants with different attack capabilities. Malware in this family uses a basic IRC network for communication between the bots and injects malicious code by exploiting a D-Link router authentication bug along with dictionary attacks. It primarily focuses on routers running MIPS and MIPSEL architectures.

Linux.Hydra Hydra tries to compromise routers on the network using a basic dictionary-based password attack with a constant list of passwords or using a D-Link authentication bypass exploit [13]. The dictionary attack targets the web-based login interface of D-Link routers on port 80 or SSH on port 22. Only routers using the MIPS architecture are targeted. The communication of the bot with the botnet is through agents using IRC communication channels. A C&C IRC server is set up by the agent, and the bot communicates with it using the IRC channel. The bot also uses wget to download the malicious binary from a link provided by the agent. Communication on the IRC channel takes place using TCP on port 6667 or nearby port numbers (6660–6669 and 7000). The agent sends commands on the channel when an attack is to be launched and for other information. The communication between the bot and the agent is, otherwise, limited. The agents can issue commands to launch a UDP flood or a TCP SYN flood attack against a victim IP also specified in the command.

PsyB0t It also uses a password dictionary attack like Linux.Hydra. The targeted routers are the ones running the MIPS or MIPSEL architectures. The botnet uses the IRC network for communication like Hydra. The difference is that the messages are encrypted and authenticated using SSL. It also tries to brute force the router password, executing shell commands, and access to other services, like MySQL, PHP MyAdmin, FTP servers, SMB shares, on the network. The agents issue commands to launch UDP or ICMP flood attacks.

Table 1 DDoS families

Family	Injection	Communication	Attack
1	–Dictionary attack (basic default passwords) –D-Link router authentication bug –Targets MIPS/MIPSEL architecture	Simple IRC network	–Direct attacks –Basic flooding attacks (UDP, TCP SYN, ACK Floods)
2	–Dictionary attacks on SSH/Telnet/ web interfaces –Targets arm, ppc, x 86 architectures also	Simple IRC network	–Direct attacks –HTTP layer 7 attacks and TCP Xmas attacks also
3	–Dictionary brute force attacks –Targets Windows/Linux systems also	Custom agent-bot handler	–Basic direct flood attacks
4	–Dictionary attacks on telnet/ssh	Heavily modified IRC network	–Basic direct flood and HTTP layer 7 attacks
5	–Dictionary attacks on telnet/ssh	Minimal peer-to-peer network only ICMP pings for enumeration)	None
6	–php vulnerability (CVE-2012-1823)	Custom channel using TCP port 58455	Crypto-mining
7	–Shellshock vulnerability –Vulnerable SOHO devices –Dictionary attacks	Custom agent-bot handler	–DNS amplification/reflection attacks
8	–Dictionary attacks, Device Specific Exploits –Drive by download from other malware	Custom HTTP-based agent-bot handler	–Basic flooding attacks –Cloudflare/Sucuri bypass techniques using Javascript
9	–Dictionary attacks on telnet/ssh	Custom agent-bot handler/modified IRC network	–Wide range of reflection/amplification floods –Stealing data from SQL databases on the network
10	–Dictionary attacks	Custom agent-bot handler	Sophisticated application-level amplification attacks

Chuck Norris It targets routers running the MIPS and MIPSEL architectures. It uses the same techniques of injection as the previous two malware. Again, an IRC-based communication network is used. Additionally, the malware tries to download a small SSH server/client that is compiled specifically for the MIPS architecture and uses it for its agent to communicate. UDP, SYN, and ACK flood attacks are deployed by the bots.

Family 2 This malware family is a more sophisticated version of Family 1, but it targets different architectures, namely ARM, PPC, and x86.

Aidra, LightAidra, Zendran The injection method is primarily password dictionary attacks on the router's SSH port or targets the web interface of the device running on port 8080/80. This malware targets more architectures including MIPS, ARM, and PPC. The attacker uses an IRC-based communication network. It only performs basic SYN and ACK flood attacks.

Tsunami/Kaiten This malware modifies the DNS settings of the host. It sets the name servers to Open DNS addresses and then blocks a range of ports from 22 to 80. It deploys traditional SYN, UDP, PUSH, and ACK flood attacks along with more sophisticated ones like HTTP layer 7 flood and TCP XMAS attacks.

Family 3 This malware family is the first one to employ a custom agent-handler botnet [13] rather than an IRC network.

Spike/Dofloo/MrBlack/Wrkatk/Sotdas/AES.DDoS It uses dictionary brute force attacks and targets various architectures, like MIPS and ARM, and also attacks Windows and Linux systems. It uses a custom agent-bot network architecture using encrypted SSL connections. The bot also periodically sends information about its computing power, so that the agent can throttle or increase the rate of attacks. This makes the attacks less conspicuous. It can launch SYN, UDP, DNS query, and GET flood attacks.

Family 4 This malware family uses a heavily modified IRC network for its botnet communications. It does not use any known IRC servers. It is also the first malware that targets devices with SPARK architectures. This family is also the first one to use telnet and ssh password brute forcing for propagation in the network, paving the way for more sophisticated malware like Mirai.

BashLite, Lizkebab, Torlus, Gafgyt It compromises devices by using dictionary attack on various services, like telnet and ssh. It uses a heavily modified IRC network architecture and is completely independent of the other IRC servers. It deploys simple UDP, SYN and ACK flood attacks.

Remaiten, KTN-RM This malware is a more powerful version of BashLite and employs a range of attacks including HTTP layer 7 floods. It is able to adapt to different architectures and devices and accordingly chooses its attacks. It is the first malware to be controlled using a custom IRC channel rather than using any existing IRC channels.

Family 5 This was the first malware, which after the infection removes other malware like Aidra and LightAidra from the device. It also blocks the usual communication ports. The important characteristic of this family is that it contains no code for the attack, but rather the goal is to spread as much as possible. The only code present in the current version of such malware is code for subverting other competing malware and maintaining communication with the bot master. In this family, the ex-filtration step is missing or "yet to happen" due to which detection is difficult.

Carna This botnet was created to build a "census" of the Internet in 2012 by using a password dictionary comprising of default or no passwords. Its goal was to remove other malware; it paved the way for the trend of other malware to remove existing malware from the device in the lateral movement step.

Linux.Wifatch It works like Carna, but it also closes the telnet port along with removing other existing malware on the infected devices. It shows a message "Telnet has been closed to avoid further infection of the device, disable telnet, change the passwords and /or update the firmware."

Hajime This is the most recent and powerful malware of this family. It uses a peer-to-peer network for its botnet communication and contains no code for the attacks. However, the bots have constant communication with the bot master, and any targeted attack code can be uploaded. Such dormant bots wait for the bot master to upload some attack code as needed when an attack is planned.

Family 6 This family was the first to perform other malicious activities rather than just flooding attacks using bots.

Linux.Darlloz This was the first malware to use bots to mine cryptocurrencies, like Mincoin and Dogecoin [38]. It exploits a php vulnerability (CVE-2012-1823) for infection and listens on TCP port 58455 for commands.

Family 7 The main targets of this malware are vulnerable SOHO (smart office and home) devices. It uses the shellshock vulnerability [35]. This family was also the first to use DNS reflection/amplification attacks [1].

Elknot, BillGates This malware uses the powerful ICMP, TCP, UDP, SYN, and HTTP layer 7 floods along with DNS reflection/amplification attacks on Linux and Windows platforms. The infection methods are the shellshock vulnerability and open ssh ports with weak passwords.

Xor.DDoS It is variant of Elknot but only targets Linux systems running x86 or ARM architectures with open ssh ports having weak passwords.

Family 8 This malware is the first that attempted to subvert DDoS prevention techniques, like Cloudflare and Sucuri [46]. This malware also uses obfuscation techniques to prevent reverse engineering of the malware.

Luabot It targets Linux systems and is written in the Lua programming language. It uses a custom HTTP-based command and control network. It contains malicious

JavaScript and also an integrated V7 embedded JavaScript engine. The goal of the JavaScript is to find the actual IP address of the victim behind a Sucuri or a Cloudflare proxy server.

Reaper Like Luabot, it contains a Lua execution environment and embeds nearly 100 open DNS resolvers that are later used to conduct DNS amplification attacks. It uses a collection of nine vulnerabilities on the D-Link 1, D-Link 2, Netgear 1, Netgear 2, Linksys, GoAhead, JAWS, Vacron, and AVTECH devices [10]. It has not been used in an active attack till now, but an analysis has shown that it has infected 2 million IoT devices.

IoTroop This is the latest variant of this family and is highly sophisticated. It has a vulnerability scanning functionality and a Lua execution environment. There is a high chance that this malware is state sponsored because of its sophistication and extensive vulnerability scanning techniques. It has a HTTP-based agent-handler architecture with the C&C server coded in php. Again, like in family 5, the malware does not contain any flooding attack code, but such code can be uploaded easily by the bot master when required.

Family 9 This is currently the most sophisticated and dangerous malware family in that it makes possible to create massive botnets. It uses password dictionary attacks for infection and borrows the spreading techniques used by family 4. It can perform a variety of attacks including sophisticated HTTP layer 7 floods. It essentially amalgamates all the useful techniques used by previous malware families.

Qbot This was the first proof-of-concept telnet-based botnet and first predecessor to Mirai. It has a simple open-sourced design and requires only two Linux servers for its implementation.

Mirai It uses the same injection techniques as BashLite. It has 60 hard-coded passwords it uses as its dictionary. Mirai stops other users and malware from accessing the device and drops communication from telnet (port 23), ssh (port 22), and web interface (port 80, 8080). It also identifies and compromises database services like MySQL and Microsoft SQL running in the network to create new admin "phpminds" with the password "phpgodwith" allowing the hackers to steal the database content. It is able to deploy a wide range of DDoS attacks, based on different protocols (e.g., TCP, UDP, and HTTP).

IRCTelnet, NewAidra It combines the dictionary attacks of Mirai with the IRC communication architecture of family 4. One unique characteristic of this malware is that it gets removed if the device is rebooted, but it can infect the device again if the injection vulnerability is not fixed.

Family 10 This is the latest variant of DDoS malware and uses sophisticated application layer amplification attacks like exploitation of the "memcached" service.

Memcrashed Memcrashed was the most devastating DDoS attack at least in terms of bandwidth with a peak of 1.35 Tbps. It used an amplification attack using misconfigured "memcached" servers. The service uses the UDP protocol and is accessible over the Internet. This resulted in an amplification factor of around 51000 in terms of bandwidth [40].

4 IoT Malware Analysis and Classification

In this section, we report our analysis on IoT malware. We collected different variants of IoT malware and built a comprehensive dataset using 3973 malware samples from the most popular malware families: Zorro, Gayfgt, Mirai, Hajime, IoTReaper, Bashlite, nttpd, linux.wifatch, etc. The malware samples were collected from IoTPOT [52], VirusTotal [63], and Open Malware [42]. These malware executables are compiled for different CPU architectures and endianness. The collected malware executables are classified according to different device architectures in Table 2. 2572 of the samples are compiled for little endian processors, while 1421 of them are for big endian processors.

System Call Analysis For analyzing these malware samples, we executed them in a safe environment according to their CPU architectures and observed the system calls made by the malware. Specifically, we focused on 34 system calls as shown in Table 3. The idea was to identify the goals of the malware by analyzing the number and duration of system calls of each type made by the malware. In our analysis, we focused on 4 parameters for each system call: number of calls made (#.1), percentage of time taken (#.2), total time taken (#.3), and average time taken per each call (#.4).

Malware Classification Using the system call analysis, we can then classify these malware in the three categories defined before: passive attacks, service degradation attacks, and DDoS attacks.

Table 2 Malware executables' breakdown according to CPU architecture

Architecture	Number of Malware samples
Mips	935
Arm	912
x86	576
Sparc	299
Renesas/SuperH	310
x86_x64	294
MC68000	294
PowerPC	353
Other	26

Table 3 Types of system calls

System call	Description
connect	Initiate a connection on a socket
_newselect	Synchronous I/O multiplexing
close	Close a file descriptor
nanosleep	High-resolution sleep
fcntl64	Manipulate file descriptor
socket	Create an end point socket for communication
rt_sigprocmask	Examine and change blocked signals
getsockopt	Get and set options on sockets
read	Read from a file descriptor
open	Open and possibly create a file
execve	Execute program
chdir	Change working directory
access	Check user's permissions for a file
brk	Change data segment size
ioctl	Manipulates the underlying device parameters of special files
setsid	Creates a session and sets the process group ID
munmap	Map or unmap files or devices into memory
wait64	Wait for process to change state
clone	Create a child process
uname	Get name and information about current kernel
mprotect	Set protection on a region of memory
prctl	Operations on a process
rt_sigaction	Examine and change a signal action
ugetrlimit	Get/set resource limits
mmap2	Map or unmap files or devices into memory
fstat64	Get file status
getuid32	Returns the real user ID of the calling process
getgid32	Returns the real group ID of the calling process
geteuid32	Returns the effective user ID of the calling process
getegid32	Returns the effective group ID of the calling process
madvise	Give advice about use of memory
set_thread_area	Set a GDT entry for thread-local storage
get_tid_address	Set pointer to thread ID
prlimit64	Get/set resource limits

Category 1 For passive attacks, the attacker uses socket system calls to establish a communication channel between the attacker and the victim to transfer data. The connect system call is then issued to use the connection to send data. The socket and connect system calls are common to all malware categories as the attacker needs to communicate with the victim. However, the amount of connect system calls made in passive data sniffing as compared to DDoS is much less. The most distinguishing system calls made by category 1 attacks are file operations, that is, read, open, and

close, and information gathering operations, that is, uname, access, and chdir. It is important to note that all these system calls are passive in nature and hard to detect as they do not modify any data on the device.

Category 2 For service degradation attacks, generally, the attacker does not require any communication channels with the victim. So, the connect and socket system calls are not required as in the other categories. Here, the most significant system calls are active system manipulation operations: nanosleep, fcntl64, rt_sigprocmask, getsockopt, brk, ioctl, munmap, clone, mprotect, and set_thread_area. Most of these malware use execve system to run malicious binaries like rootkits and prtcl to change the name of the malicious process (using PR_SET_NAME request) to a benign service name like notepad, sshd, telnet, etc. Most of these system calls are active and can be detected more easily than in the attacks in category 1. However, since these system calls are commonly used by benign applications, differentiating between benign applications and malware is difficult and results in false positives during detection.

Category 3 For DDoS attacks, once a communication channel is established between the victim and the attacker using socket and connect system calls, the malware generally stays dormant. Some of the obfuscation and lateral movement techniques are common to previous categories in some malware as the goal of the attacker is to stay as inconspicuous as possible until ex-filtration or launching the attack.

5 AI-Based IDS Solutions

In this section, we survey AI-based IDSs developed for IoT networks. We analyze how detection models are built and how they would benefit from our classification scheme.

IDSs for IoT networks typically build detection models based on network flows, device system data, or user/app access/usage patterns. A network flow is defined for each (*SourceIP:SourcePort:Destination IP:Destination Port*) pair, and for every such flow, the following parameters are used to build flow-based models: data sizes, data rate, IP options, protocols (application/link/network layers), specific protocol parameters for ICMP, UDP, TCP, HTTP, etc., and cloud service APIs (if used) [22, 44, 54]. Some IDSs focus on the system features of the IoT device through process and system call information to build detection models [48, 64]. Recently, approaches have been proposed that build static or dynamic smart app models [6, 7, 18, 25, 47, 59, 64]. The models are generated using a combination of device capabilities, user prompts, app permissions, and event subscriptions.

Such IDS models can be categorized into three categories: signature based, anomaly based, and state-full protocol analysis. Signature-based models are built using supervised learning from predefined or known attacks from which classifiers are trained to detect such attacks. Classical machine learning techniques like decision trees [33], SVMs [53], and random forests [16] have been used in the past to build such classifiers. Recently, deep neural networks (DNNs)-based models [15, 30, 66] have become very accurate and can deal with large datasets of known attacks. By contrast, anomaly-based techniques observe benign behavior to build baseline models and then detect anomalies against these baselines to detect attacks. Classical machine learning techniques, such as clustering [27], are being replaced by techniques, like DNNs [62], for building anomaly-based models. State-full protocol analyses involve a dynamic temporal analysis of device behavior to build time series data or graphs. In traditional machine learning settings, some form of forward symbolic execution [28] is used to explore all feasible paths in the time series data or built graphs. Then, either anomaly detection or known attack signature matching is performed to detect attacks. With the growing size of data, classical techniques are infeasible, and AI techniques like recurrent neural networks (RNNs) and long short-term memory (LSTM) are used to build such state-full detection models [17, 29]. Recently, deep reinforcement learning (DRL) techniques are being used to explore the environment freely and learn better detection models from known attacks [39].

The accuracy and efficiency of the AI-based detection models mentioned above depend on the data collected from known attacks and benign IoT device behavior. Most of these models, however, are built for traditional networks where the kill chain of the attack is fairly well understood. In most cases, the detection models are built to be deployed at the potential infection points or for detecting the initial steps of the kill chain. However, for IoT networks, such a strategy for building models results in inefficient models with low accuracy and higher false positives. Our classification strategy attempts to address this issue by providing a clear classification of IoT attacks and malware based on the "goals of the attacker." The data collected can thus be preprocessed and analyzed based on the last three steps of the kill chain and result in more efficient and accurate models.

6 Conclusion

In this chapter, we have provided a comprehensive survey of IoT attacks and classified them into three categories according to the goals of the attacker. We have analyzed these attacks in terms of a kill chain and provided a means of building more accurate AI-based models for detection. As part of our future work, we plan to use this classification and analyze system call behavior of IoT malware further to build efficient AI-based detection models.

References

1. Anagnostopoulos, M., Kambourakis, G., Kopanos, P., Louloudakis, G., Gritzalis, S.: DNS amplification attack revisited. Comput. Secur. **39**, 475–485 (2013)
2. Angrishi, K.: Turning internet of things (IoT) into internet of vulnerabilities (IoV): IoT botnets. Preprint (2017). arXiv:1702.03681
3. Apthorpe, N., Reisman, D., Sundaresan, S., Narayanan, A., Feamster, N.: Spying on the smart home: Privacy attacks and defenses on encrypted IoT traffic. Preprint (2017). arXiv:1708.05044
4. Borella, M.S.: Source models of network game traffic. Computer Communications **23**(4), 403–410 (2000)
5. Cao, T., Shen, P., Bertino, E.: Cryptanalysis of some rfid authentication protocols. J. Commun. **3**(7), 20–27 (2008)
6. Celik, Z.B., McDaniel, P., Tan, G.: Soteria: Automated IoT safety and security analysis. In: 2018 USENIX Annual Technical Conference, pp. 147–158 (2018)
7. Celik, Z.B., Tan, G., McDaniel, P.D.: IoTGuard: Dynamic enforcement of security and safety policy in commodity IoT. In: NDSS (2019)
8. Choi, S.J., Kwak, J.: A study on reduction of DDoS amplification attacks in the UDP-based CLDAP protocol. In: 2017 4th International Conference on Computer Applications and Information Processing Technology (CAIPT), pp. 1–4. IEEE (2017)
9. Cimpanu, C.: Brickerbot author claims he bricked two million devices. Bleeping Computer, April (2017)
10. Cimpanu, C.: A gigantic IoT botnet has grown in the shadows in the past month (Oct 2017). https://www.bleepingcomputer.com/news/security/a-gigantic-iot-botnet-has-grown-in-the-shadows-in-the-past-month/
11. Cobb, S.: Rot: Ransomware of things (2017)
12. Coleman, C.: Addressing the cyber kill chain: Full Gartner research report and looking glass perspectives (2016)
13. De Donno, M., Dragoni, N., Giaretta, A., Spognardi, A.: DDoS-capable IoT malwares: Comparative analysis and mirai investigation. Secur. Commun. Networks **2018** (2018). https://doi.org/10.1155/2018/7178164
14. Devices, C., Health, R.: Safety communications - firmware update to address cybersecurity vulnerabilities identified in Abbott's (formerly St. Jude Medical's) implantable cardiac pacemakers: FDA safety communication. https://www.fda.gov/MedicalDevices/Safety/AlertsandNotices/ucm573669.htm
15. Diro, A.A., Chilamkurti, N.: Distributed attack detection scheme using deep learning approach for internet of things. Future Gener. Comput. Syst. **82**, 761–768 (2018)
16. Farnaaz, N., Jabbar, M.: Random forest modeling for network intrusion detection system. Procedia Comput. Sci. **89**(1), 213–217 (2016)
17. Feng, C., Li, T., Chana, D.: Multi-level anomaly detection in industrial control systems via package signatures and LSTM networks. In: 2017 47th Annual IEEE/IFIP International Conference on Dependable Systems and Networks (DSN), pp. 261–272. IEEE (2017)
18. Fernandes, E., Jung, J., Prakash, A.: Security analysis of emerging smart home applications. In: Proceedings of the 2016 IEEE Symposium on Security and Privacy, pp. 636–654. IEEE (2016)
19. Genge, B., Enăchescu, C.: Shovat: Shodan-based vulnerability assessment tool for internet-facing services. Secur. Commun. Networks **9**(15), 2696–2714 (2016)
20. Ghafoor, I., Jattala, I., Durrani, S., Tahir, C.M.: Analysis of OpenSSL Heartbleed vulnerability for embedded systems. In: 17th IEEE International Multi Topic Conference 2014, pp. 314–319. IEEE (2014)
21. Goland, Y.Y., Cai, T., Leach, P., Gu, Y., Albright, S.: Simple service discovery protocol (1999)
22. Habibi, J., Midi, D., Mudgerikar, A., Bertino, E.: Heimdall: Mitigating the internet of insecure things. IEEE Internet Things J. **4**(4), 968–978 (2017)

23. Ho, G., Leung, D., Mishra, P., Hosseini, A., Song, D., Wagner, D.: Smart locks: Lessons for securing commodity internet of things devices. In: Proceedings of the 11th ACM on Asia Conference on Computer and Communications Security, pp. 461–472. ACM (2016)
24. Iotroop botnet: The full investigation (Feb 2018), https://research.checkpoint.com/iotroop-botnet-full-investigation/
25. Jia, Y.J., Chen, Q.A., Wang, S., Rahmati, A., Fernandes, E., Mao, Z.M., Prakash, A., Unviersity, S.J.: Contexlot: Towards providing contextual integrity to amplified IoT platforms. In: NDSS (2017)
26. Kenin, S.: Brickerbot analysis (Dec 2017). =https://www.trustwave.com/Resources/SpiderLabs-Blog/BrickerBot-mod-plaintext-Analysis
27. Khan, L., Awad, M., Thuraisingham, B.: A new intrusion detection system using support vector machines and hierarchical clustering. VLDB J. **16**(4), 507–521 (2007)
28. Khurshid, S., Păsăreanu, C.S., Visser, W.: Generalized symbolic execution for model checking and testing. In: International Conference on Tools and Algorithms for the Construction and Analysis of Systems, pp. 553–568. Springer (2003)
29. Kim, G., Yi, H., Lee, J., Paek, Y., Yoon, S.: LSTM-based system-call language modeling and robust ensemble method for designing host-based intrusion detection systems. Preprint (2016). arXiv:1611.01726
30. Kolosnjaji, B., Zarras, A., Webster, G., Eckert, C.: Deep learning for classification of malware system call sequences. In: Australasian Joint Conference on Artificial Intelligence, pp. 137–149. Springer (2016)
31. Krebs, B.: Krebs on security. https://www.krebsonsecurity.com/2017/10/fear-the-reaper-or-reaper-madness
32. Krebs, B.: Krebs on security. https://krebsonsecurity.com/2017/10/reaper-calm-before-the-iot-security-storm/
33. Kruegel, C., Toth, T.: Using decision trees to improve signature-based intrusion detection. In: International Workshop on Recent Advances in Intrusion Detection, pp. 173–191. Springer (2003)
34. Lee, J., Stanley, M., Spanias, A., Tepedelenlioglu, C.: Integrating machine learning in embedded sensor systems for internet-of-things applications. In: 2016 IEEE International Symposium on Signal Processing and Information Technology (ISSPIT), pp. 290–294. IEEE (2016)
35. Leyden, J.: Patch bash now: "shellshock" bug blasts OS X, Linux systems wide open (2014)
36. Li, H., Ota, K., Dong, M.: Learning IoT in edge: Deep learning for the internet of things with edge computing. IEEE Network **32**(1), 96–101 (2018)
37. Li, T., Wang, G.: Security analysis of two ultra-lightweight rfid authentication protocols. In: IFIP International Information Security Conference, pp. 109–120. Springer (2007)
38. Linux.darlloz.: =https://www.symantec.com/security-response/writeup.jsp?docid=2013-112710-1612-99&tabid=2
39. Lopez-Martin, M., Carro, B., Sanchez-Esguevillas, A.: Application of deep reinforcement learning to intrusion detection for supervised problems. Expert Syst. Appl. **141**, 112963 (2020)
40. Majkowski, M.: Memcrashed - major amplification attacks from UDP port 11211. blog.cloudflare.com/memcrashed-major-amplification-attacks-from-port-11211, accessed: 2018-03-07
41. Malkin, G., et al.: Rip version 2. Tech. rep., STD 56, RFC 2453, November (1998)
42. Malware, O.: https://openmalware.org
43. Martin, L.: Cyber kill chain®. URL: https://www.cyber.com. lockheedmartin. com/hubfs/Gaining the Advantage Cyber Kill Chain. pdf (2014)
44. Midi, D., Rullo, A., Mudgerikar, A., Bertino, E.: Kalis–a system for knowledge-driven adaptable intrusion detection for the internet of things. In: 2017 IEEE 37th International Conference on Distributed Computing Systems (ICDCS), pp. 656–666. IEEE (2017)
45. Mills, D.L.: Internet time synchronization: the network time protocol. IEEE Trans. Commun. **39**(10), 1482–1493 (1991)

46. Mmd-0057-2016 - linux/luabot - iot botnet as service · malwaremustdie!: http://blog.malwaremustdie.org/2016/09/mmd-0057-2016-new-elf-botnet-linuxluabot.html

47. Mohsin, M., Anwar, Z., Husari, G., Al-Shaer, E., Rahman, M.A.: IoTSAT: A formal framework for security analysis of the internet of things (IoT). In: 2016 IEEE Conference on Communications and Network Security (CNS), pp. 180–188. IEEE (2016)

48. Mudgerikar, A., Sharma, P., Bertino, E.: E-Spion: A system-level intrusion detection system for IoT devices. In: Proceedings of the 2019 ACM Asia Conference on Computer and Communications Security, pp. 493–500 (2019)

49. Muñoz, R., Vilalta, R., Yoshikane, N., Casellas, R., Martínez, R., Tsuritani, T., Morita, I.: Integration of IoT, transport SDN, and edge/cloud computing for dynamic distribution of IoT analytics and efficient use of network resources. J. Lightwave Technol. **36**(7), 1420–1428 (2018)

50. News, C.: Car hacked on 60 minutes (Feb 2015). https://www.cbsnews.com/news/car-hacked-on-60-minutes/

51. Nurse, J.R., Erola, A., Agrafiotis, I., Goldsmith, M., Creese, S.: Smart insiders: exploring the threat from insiders using the internet-of-things. In: 2015 International Workshop on Secure Internet of Things (SIoT), pp. 5–14. IEEE (2015)

52. Pa, Y.M.P., Suzuki, S., Yoshioka, K., Matsumoto, T., Kasama, T., Rossow, C.: IoTPOT: analysing the rise of IoT compromises. EMU **9**, 1 (2015)

53. Rao, X., Dong, C.X., Yang, S.Q.: An intrusion detection system based on support vector machine. J. Software **14**(4), 798–803 (2003)

54. Raza, S., Wallgren, L., Voigt, T.: Svelte: Real-time intrusion detection in the internet of things. Adhoc Networks **11**(8), 2661–2674 (2013)

55. Rouse, M.: What is IoT (internet of things) and how does it work? (Feb 2020). https://internetofthingsagenda.techtarget.com/definition/Internet-of-Things-IoT

56. Sarkar, P.G., Fitzgerald, S.: Attacks on SSL a comprehensive study of beast, crime, time, breach, lucky 13 & rc4 biases. Internet: https://www.isecpartners.com/media/106031/ssl_attacks_survey. pdf [June, 2014] (2013)

57. Srinivasan, R.: RPC: Remote procedure call protocol specification version 2 (1995)

58. Strom, B.E., Applebaum, A., Miller, D.P., Nickels, K.C., Pennington, A.G., Thomas, C.B.: Mitre att&ck: Design and philosophy. Technical report (2018)

59. Tian, Y., Zhang, N., Lin, Y.H., Wang, X., Ur, B., Guo, X., Tague, P.: Smartauth: User-centered authorization for the internet of things. In: Proceedings of the 26th USENIX Security Symposium, pp. 361–378 (2017)

60. Valente, J., Cardenas, A.: Security and privacy in smart toys. In: Proceedings of the 2017 Workshop on Internet of Things Security and Privacy, IoT S&P@CCS, Dallas, TX, USA, November 03, 2017, pp. 19–24 (2017)

61. Valeriano, B., Habel, P.: Who are the enemies? the visual framing of enemies in digital games. Int. Stud. Rev. **18**(3), 462–486 (2016)

62. Van, N.T., Thinh, T.N., et al.: An anomaly-based network intrusion detection system using deep learning. In: 2017 International Conference on System Science and Engineering (ICSSE), pp. 210–214. IEEE (2017)

63. VirusTotal: https://www.virustotal.com

64. Wang, Q., Hassan, W.U., Bates, A., Gunter, C.: Fear and logging in the internet of things. In: Network and Distributed Systems Symposium (2018)

65. Ward, M.: Smart meters can be hacked to cut power bills (Oct 2014). http://www.bbc.com/news/technology-29643276

66. Xiao, L., Wan, X., Lu, X., Zhang, Y., Wu, D.: IoT security techniques based on machine learning: How do IoT devices use AI to enhance security? IEEE Signal Process. Mag. **35**(5), 41–49 (2018)

67. Zeifman, I., Bekerman, D., Herzberg, B.: Breaking down mirai: An IoT DDoS botnet analysis (2016). Imperva. Source: https://www.incapsula.com/blog/malware-analysis-mirai-ddos-botnet.html

Machine Learning-Based Online Source Identification for Image Forensics

Yonggang Huang, Lei Pan, Wei Luo, Yahui Han, and Jun Zhang

1 Introduction

Nowadays, digital images provide key information of cyber crimes in digital foren-sics. Image source identification is an important branch of image forensics because it is used to associate an image with the acquisition device [34]. Representative examples include the identification of source device of a specific evidence image at a court of law and the verification of the real owner of images during copyright disputes.

Presently, three categories of approaches are commonly employed to identify image source. These methods are *image metadata based*, *watermark based*, or *feature based*. The image metadata based approach directly investigates the image source information embedding in the EXIF (Exchangeable Image File) header [4], which is apparent and simple. However, the EXIF header can be easily modified in real life, making this approach unreliable. The watermark based approach [12, 39] relies on the watermarks injected into the images with the source related information. The watermark is difficult to be tampered with, so that this approach is robust. Unfortunately, camera producers and vendors are reluctant to apply watermarks in the large scale due to the huge costs by introducing the additional

Y. Huang (✉) · Y. Han
School of Computer Science and Technology, Beijing Institute of Technology, Beijing, China
e-mail: 3220190803@bit.edu.cn

L. Pan · W. Luo
School of Information Technology, Deakin University, Geelong, VIC, Australia
e-mail: l.pan@deakin.edu.au; wei.luo@deakin.edu.au

J. Zhang
School of Software and Electrical Engineering, Swinburne University of Technology, Hawthorn, VIC, Australia
e-mail: junzhang@swin.edu.au

watermark injection module. Recently, the feature based approach has thrived [1, 5, 6, 26, 33] because it is relatively easy to operate during the image acquisition process. And the intrinsic features are extracted from hardware artifacts or from software-related fingerprints. Hence, the device identification process becomes a machine learning-based classification problem that can be solved with the statistical learning tools. This paper focuses on the feature based approach.

Existing statistical learning solutions for the feature based image source identification consist of two categories—*closed set oriented* and *open set oriented*. The closed set oriented solutions assume that the identification task is performed on a closed dataset. That is, we collect and categorize the images from many known camera models before assigning the target images to any given camera models. Therefore, image source identification becomes a K-class classification problem, where K denotes the number of known models. Such classification problems can be solved with multi-class classifiers, such as multi-class Support Vector Machine (SVM) [10]. Many existing solutions [1, 5, 6, 26, 33] belong to the closed set oriented category. However, the open set oriented methods aim to deal with a challenging but realistic situation. That is, the identification process is carried out on open dataset, and the target images may be taken by multiple unknown camera models. In this scenario, the challenge is how to identify images of the unknown models while correctly sorting the images shot by the known camera models [11, 18, 21, 22, 40]. A few publications [11, 18, 21, 22, 40] have attempted to address this issue with limited success, and this paper belongs to this category with better outcomes.

The related work assumes that the identification process is conducted in an offline manner. That is, all the testing images are given at once, and the unknown is constant in the open set scenario. In this paper, we consider the online image source identification. That is, the image source identification is deployed as an online service, and the testing images are arriving continuously as an image stream. In such an online identification situation, the images shot by unknown camera models are emerging dynamically alongside with the image stream. For example, during a given time duration t, the image stream contains the images shot by K known camera models and by N unknown camera models. In the time duration $(t + 1)$, the images from another unknown camera model become present in the stream. In fact, new models of digital cameras and new lenses are continuously introduced to the market.

This work investigates the online issue for open set oriented image source identification. This paper aims to develop a new online image source identification scheme capable of effectively detecting dynamically emerging unknown models. The main contributions of this paper are manifold and summarized as follows:

- To the best of our knowledge, this paper pioneers in the identification and solution of unknown camera models for the online image source identification task.
- A novel scheme, namely, Online image Source Identification with Unknown camera model (OSIU), is proposed in this paper. OSIU can effectively identify

the images shot by unknown camera models via a 3-step process including unknown sample triage, unknown image discovery, and $(K + 1)$-class classification.

- A novel unknown sample triage method is developed to detect the presence of any image shot by any camera models previously unknown.
- A new unknown image discovery algorithm is introduced to verify the source of the samples shot by unknown camera models. Moreover, a novel parameter optimization method is also developed to improve the accuracy of the procedure of unknown image discovery.

Due to the complexity of the underlying algorithms, our unknown image discovery spends significantly longer time than our unknown sample triage. However, the former procedure is more accurate than the latter. In order to balance the accuracy and computation cost, our scheme OSIU executes the unknown sample triage to identify the unknown images in an image cache before executing the unknown image discovery.

The rest of the paper is organized as follows: Sect. 2 reviews the related work. Section 3 presents OSIU as a new image source identification scheme. Section 4 reports our experiments and results. Section 5 concludes this paper.

2 Related Work

Recent research attention has been increasingly paid to the feature based image source identification solutions [1, 5, 6, 26, 33]. Most of the existing research focus on feature engineering, and others rely on statistics based detection methods. However, to the best of our knowledge, the problem of identifying images shot by unknown camera models as part of online image source identification has not been well addressed in previous work.

2.1 Features Engineering for Image Source Identification

Research efforts in this category commit capturing the key features as the representation of the intrinsic fingerprints generated during the image acquisition process, such as hand-crafted features and deep learning features.

In general, two sets of hand-crafted features are available for image source identification depending on the sources of the features, including hardware- and software-related fingerprints. Regarding the hardware fingerprints, there are four categories, that is, pattern noise [24, 25, 28–30, 35, 36], lens radial distortion [8, 9, 32], chromatic aberration [38], and sensor dust [13, 14]; and regarding the software-related fingerprints, there are two categories, that is, image-related features [26] and artifacts introduced by color filter array [1]. According to Kharrazi et al. [26], the color filter array configuration/demosaicing algorithm and the color processing/transformation significantly alter the output image. This theory is proven by the 34 purposely designed features to capture the underlying color characteristics of different cameras. In [17], camera models were reliably identified by investigating the patterns of the perfect pixels or defective pixels. In [14], the sensor dust traces formed by the location and shape of dust specks in front of the imaging sensor and their persistence were used to identify digital single lens reflex cameras. In [9], the intrinsic lens radial distortion was used for image source identification because of the ubiquitous existence of radial distortions of the spherical surfaced lenses used in almost all digital cameras. Recently, photo-response nonuniformity noise (PRNU) has become a new type of fingerprint for camera identification. In [30], original images were denoised through a wavelet denoising filter to obtain the PRNU; subsequently, the reference PRNU of the camera was obtained by averaging the PRNU of multiple reference images; lastly, the image source identification was measured according to the correction between PRNU of an image and the camera reference PRNU. Some variants of this procedure can be found in literatures [25, 29, 35]. For example, Li et al. [29] found that the extracted PRNU can be affected by the color filter array. To mitigate the effects of the color filter array, each color channel can be further decomposed into four sub-images each of which can be used to extract a sub-PRNU. At last, the four sub-PRNU components are aggregated into the final PRNU that is significantly more robust than the single PRNU extracted from the original image. Last but not the least, Pan and Trepanic [31] successfully applied PRNU signatures to recognize the Facebook images shot by the iPhone models.

Instead of imposing a predetermined analytical model, the deep learning features are data-driven and learned directly from the data itself. In [37], a 7-layered Convolutional Neural Networks (CNNs) framework was proposed for source camera identification with the following configuration—the first layer is a filter layer; the second, third, and fourth layers are three convolutional layers; the last three layers are three fully connected layers. In [2, 3], the CNN network structure was modified with four convolutional layers and two subsequent inner product layers. In [15], a 6-layered CNN architecture was proposed for camera model identification where there are two convolutional layers, a max pooling layer, and three fully connected layers.

2.2 Statistical Learning-Based Image Source Identification

Upon the completion of extracting the features from the images, image source identification can be regarded as a machine learning problem. The existing publications using machine learning for image source identification can be summarized in two groups, that is, *closed set oriented* and *open set oriented*.

The closed set oriented solutions assume that the identification task operates on a closed dataset. In this scenario, the target images are assumed to be shot by a known camera model. More concretely, the identification problem on a closed set can be defined as a K-class classification problem where K stands for the number of known camera models. Such classification tasks can be easily solved by training multi-class classifiers. Hence, most of the existing works [1, 5, 6, 26, 33] belong to this category.

On the contrary, the open set oriented methods admit the fact that target images may be taken by the camera whose models are previously unknown to us. In this situation, all images need to be correctly identified no matter whether by unknown camera models or by known camera models. Due to the difficulty of the open set oriented approach, there are only a few publications attempting to solve the problem. In [18] two SVM-based approaches were proposed, namely, one-class SVM (OSVM) and binary SVM (BSVM). In OSVM, a one-class SVM was trained to separate the images from the known models. OSVM is relatively simple so that it can be used to sort the target images quickly. The BSVM approach was developed by adopting the one-against-one [21] multi-class SVM implementation. In BSVM, all known models are regarded as K-classes, and the unknown models are treated as an additional class; therefore, the image identification problem is solved by using a $(K + 1)$-class classifier. In [40], a combined classification framework (CCF) was proposed. In CCF, a one-class SVM is trained to determine whether an image is from the known model. The one-class SVM classifier will produce binary outputs—if the given image is not known to the classifier, then it is regarded as an unknown sample; otherwise, the image is fed to a multi-class SVM classifier to determine which known model it belongs to. Furthermore, in [11], a decision boundary carving based approach (DBC) was proposed. In DBC, a binary SVM is trained with the data samples including positive samples and negative samples—the positive samples are the images taken by any known camera model, and the negative samples are the images taken by any unknown camera model. By considering the unknown camera models, the decision boundary of the SVM is gradually adjusted toward the positive class and away from the negative class to minimize the errors. More importantly, a novel scheme, namely, Source Camera Identification with Unknown model (SCIU), was proposed to address the problem of unknown camera models in [22]. SCIU explores the information of unknown camera models through two stages—unknown detection and unknown expansion before incorporating the information of the unknown camera models into the $(K + 1)$-class classification.

3 Proposed Scheme: OSIU

This section presents the details of our proposed scheme OSIU. Figure 1 depicts the system model of our scheme. Our new solution consists of three key components—unknown sample triage, unknown image discovery, and $(K + 1)$-class classification. To speed up the process, we set up an image cache to temporarily store a small number of new images to be processed; but we discard all the cached images once the cache is full.

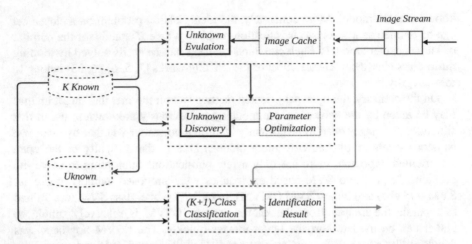

Fig. 1 The system model of OSIU (the blue line, red line, and green line indicate the data flow of unknown sample triage, unknown image discovery, and $(K + 1)$-class classification, respectively)

3.1 Unknown Sample Triage

The existing works solved the problem of detecting the images with unknown samples in an offline manner. However, such solutions do not meet the requirement of real-world image detection problems without covering the online learning scenario. Furthermore, although the information embedded in the images shot by unknown camera models can be extracted via machine learning methods as presented in Sect. 3.2, the discovery process is often computationally expensive. Thus, it is computationally infeasible to execute such resource intensive process across many image caches in real time. To reduce the overall computational cost, we propose to triage the images in the image cache. During the triage step, the images shot by any unknown model will be identified and immediately redirected to the unknown image discovery process. Our aim is to reduce the overall processing time by increasing the throughput of the entire process to meet the real-time requirement.

The basic idea of unknown sample triage is to measure the divergence between the underlying distribution of training dataset and that of the image cache. Two statistics are used to capture properties of the underlying distributions: The first statistic is derived from the class frequency distribution of camera models; the second statistic is derived based on confusion relationships of different camera models.

Statistic I: Kullback–Leibler Distance of Class Frequency Distribution of Camera Models Given the training dataset $\mathbb{T}$, which consists of images shot by K camera models denoted as $\{C_1, \cdots, C_K\}$. And we use $\mathbb{M}$ to denote the SVM model trained on the training set $\mathbb{T}$. Any image can be represented as a probability vector with the probability values predicted by the trained SVM. In this paper, we use LIBSVM [7] to implement our SVM models with the support of probability estimation via pairwise coupling [41]. Each image I can be represented as follows:

$$e(I) = [e_1(I), \cdots, e_K(I)] \leftarrow \mathsf{SVMPredict}(\mathbb{M}, I), \qquad (1)$$

where $e_i(I)$ indicates image I's predicted probability with respect to C_i.

The class frequency distribution of the training dataset $\mathbb{T}$ can be estimated via combing the probability representation of all samples, such that

$$P(\mathbb{T}) = [P_1(\mathbb{T}), \cdots, P_K(\mathbb{T})] = \left[\frac{\sum_{I \in \mathbb{T}} e_1(I)}{|\mathbb{T}|}, \cdots, \frac{\sum_{I \in \mathbb{T}} e_K(I)}{|\mathbb{T}|} \right], \qquad (2)$$

where $P_i(\mathbb{T})$ is the class frequency of model C_i within $\mathbb{T}$.

Within any time duration t, the image cache temporarily contains a set of images denoted as Φ_t. The class frequency distribution of Φ_t can be calculated as follows:

$$Q(\Phi_t) = [Q_1(\Phi_t), \cdots, Q_K(\Phi_t)] = \left[\frac{\sum_{I \in \Phi_t} e_1(I)}{|\Phi_t|}, \cdots, \frac{\sum_{I \in \Phi_t} e_K(I)}{|\Phi_t|} \right], \qquad (3)$$

where $Q_i(\Phi_t)$ is Φ_t's class frequency of model C_i.

For an image cache without images shot by any unknown camera models, its class frequency distribution should be similar to that of the training dataset. That is, $P(\mathbb{T}) \approx Q(\Phi_t)$. Conversely, if an image cache contains many images from unknown models, its class frequency distribution will be quite different to that of the training dataset. This distribution divergence is caused by misclassification of images from unknown models as known models. This hypothesis is validated through the empirical studies as shown in Fig. 2. We use the following parameters in the experiments in Sect. 4—the image cache size $s = 4000$ and the unknown information increment $u = 3$.

Figure 2 depicts the class frequency distributions of the training dataset and the image cache. The class frequency distribution of training dataset is similar to that of the image cache without images shot by unknown camera models but quite different to that of the image cache containing images shot by the 9 unknown models.

To measure the divergence between the class frequency distributions of different dataset, we use the Kullback–Leibler distance [27] as following:

$$D_{KL}(Q(\Phi_t) \| P(\mathbb{T})) = \sum_i Q_i(\Phi_t) \log \frac{Q_i(\Phi_t)}{P_i(\mathbb{T})}. \qquad (4)$$

Statistic II: Pearson's Correlation Coefficient of Confusion Matrices of Camera Models According to Eq. 1, each candidate image I can be represented as a probability vector $e(I) = [e_1(I), \cdots, e_K(I)]$. The autocorrelation matrix of $\mathbb{E}(I)$

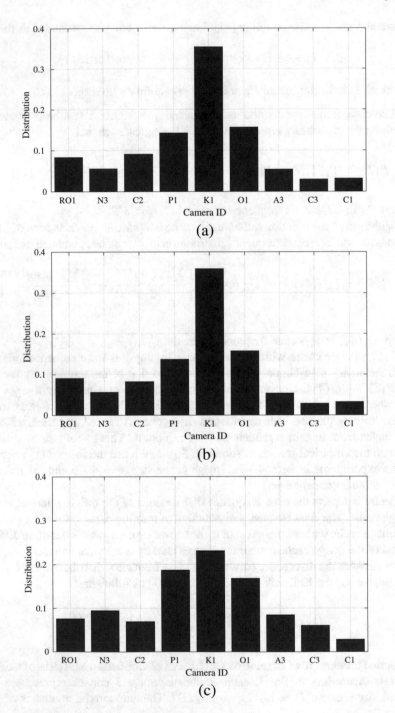

Fig. 2 Class frequency distributions of training dataset and image cache. (**a**) Training dataset without images shot by unknown camera models. (**b**) Image cache without images shot by unknown camera models. (**c**) Image cache with images shot by 9 unknown camera models

can be calculated by multiplying $e(I)^T$ and $e(I)$

$$\mathbb{E}(I) = e(I)^T \cdot e(I) = \begin{pmatrix} e_1(I)e_1(I) & e_1(I)e_2(I) & \cdots & e_1(I)e_K(I) \\ e_2(I)e_1(I) & e_2(I)e_2(I) & \cdots & e_2(I)e_K(I) \\ \vdots & \vdots & \ddots & \vdots \\ e_K(I)e_1(I) & e_K(I)e_2(I) & \cdots & e_K(I)e_K(I) \end{pmatrix}. \quad (5)$$

The autocorrelation matrix of the training dataset can be obtained by combining the autocorrelation matrices of all the samples contained in $\mathbb{T}$ as follows:

$$\Lambda(\mathbb{T}) = \frac{\sum_{I \in \mathbb{T}} e(I)^T \cdot e(I)}{|\mathbb{T}|} =$$

$$\frac{1}{|\mathbb{T}|} \begin{pmatrix} \sum_{I \in \mathbb{T}} e_1(I)e_1(I) & \sum_{I \in \mathbb{T}} e_1(I)e_2(I) & \cdots & \sum_{I \in \mathbb{T}} e_1(I)e_K(I) \\ \sum_{I \in \mathbb{T}} e_2(I)e_1(I) & \sum_{I \in \mathbb{T}} e_2(I)e_2(I) & \cdots & \sum_{I \in \mathbb{T}} e_2(I)e_K(I) \\ \vdots & \vdots & \ddots & \vdots \\ \sum_{I \in \mathbb{T}} e_K(I)e_1(I) & \sum_{I \in \mathbb{T}} e_K(I)e_2(I) & \cdots & \sum_{I \in \mathbb{T}} e_K(I)e_K(I) \end{pmatrix}. \quad (6)$$

Similarly, the autocorrelation matrix of the image cache can be obtained by combining the autocorrelation matrices of all the samples contained in Φ_t during the time duration t as follows:

$$\Gamma(\Phi_t) = \frac{\sum_{I \in \Phi_t} e(I)^T \cdot e(I)}{|\Phi_t|} =$$

$$\frac{1}{|\Phi_t|} \begin{pmatrix} \sum_{I \in \Phi_t} e_1(I)e_1(I) & \sum_{I \in \Phi_t} e_1(I)e_2(I) & \cdots & \sum_{I \in \Phi_t} e_1(I)e_K(I) \\ \sum_{I \in \Phi_t} e_2(I)e_1(I) & \sum_{I \in \Phi_t} e_2(I)e_2(I) & \cdots & \sum_{I \in \Phi_t} e_2(I)e_K(I) \\ \vdots & \vdots & \ddots & \vdots \\ \sum_{I \in \Phi_t} e_K(I)e_1(I) & \sum_{I \in \Phi_t} e_K(I)e_2(I) & \cdots & \sum_{I \in \Phi_t} e_K(I)e_K(I) \end{pmatrix}. \quad (7)$$

The elements in the above autocorrelation matrices reflect the occurrence frequency of classes. In particular, the off-diagonal elements correspond to the co-activation of different classes. Ideally, there is only one active class at each time instance. From this viewpoint, the off-diagonal elements reflect confusions among the classes. For an image cache without any images taken by unknown camera models, the autocorrelation matrix should be similar to that of the training dataset. In contrast, for an image cache containing images shot by the unknown camera models, more non-zero off-diagonal elements would appear in the autocorrelation matrix, compared to that of the training dataset. The reason is that the images shot by the unknown camera models distort the occurrence frequency of classes.

The autocorrelation matrices of training dataset and image caches are presented in Fig. 3 in color maps. It is shown that the autocorrelation matrix of the image

Fig. 3 Autocorrelation matrix of training dataset and image cache. (**a**) Training dataset without images shot by unknown camera models. (**b**) Image cache without images shot by unknown camera models. (**c**) Image cache with images shot by 9 unknown models

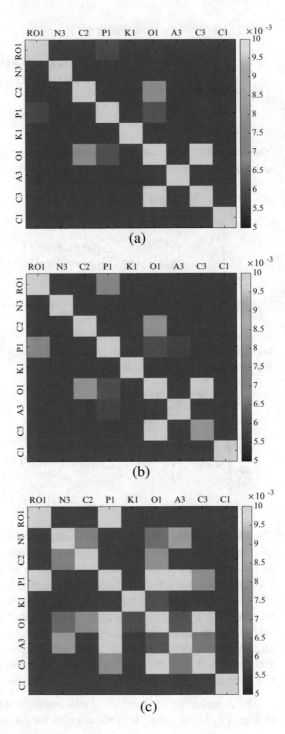

(a)

(b)

(c)

cache without unknown is similar to that of the training dataset. Comparatively, the autocorrelation matrix of the image cache with 9 unknown models has more non-zero off-diagonal elements.

To measure the similarities between the autocorrelation matrices, we use the Pearson's correlation coefficient [16] as follows:

$$Cor(\Lambda(\mathbb{T}), \Gamma(\Phi_t)) =$$
$$\frac{\sum_{i,j}(\Lambda_{i,j}(\mathbb{T}) - \bar{\Lambda}(\mathbb{T})(\Gamma_{i,j}\Phi_t) - \bar{\Gamma}(\Phi_t))}{\sqrt{\sum_{i,j}(\Lambda_{i,j}(\mathbb{T}) - \bar{\Lambda}(\mathbb{T}))^2}\sqrt{\sum_{i,j}(\Gamma_{i,j}(\Phi_t)) - \bar{\Gamma}(\Phi_t)))^2}}, \tag{8}$$

where $\Lambda_{i,j}(\mathbb{T})$ is the i-th row, j-th column element in matrix $\Lambda(\mathbb{T})$, and $\bar{\Lambda}(\mathbb{T})$ represents the arithmetic mean value of $\Lambda(\mathbb{T})$.

To measure the divergence of the training dataset and the image cache, we subtract the Pearson's correlation coefficient $Cor(\Lambda(\mathbb{T}), \Gamma(\Phi_t))$ with the Kullback–Leibler distance $D_{KL}(Q(\Phi_t)||P(\mathbb{T}))$. Hence, the divergence $S(\mathbb{T}, \Phi_t)$ is derived by

$$S(\mathbb{T}, \Phi_t) = -D_{KL}(Q(\Phi_t)||P(\mathbb{T})) + Cor(\Lambda(\mathbb{T}), \Gamma(\Phi_t)). \tag{9}$$

The divergence increment with respect to the training set and the image cache can be calculated as follows:

$$\Delta S(\mathbb{T}, \Phi_t) = |S(\mathbb{T}, \Phi_t) - S(\mathbb{T}, \Phi_d)|, \tag{10}$$

where d is the last time duration when any new sample shot by unknown camera models is detected. The default value of d is the duration of the service startup.

To flag an image cache with suspicious samples shot by unknown camera models, we compare the divergence increment $\Delta S(\mathbb{T}, \Phi_t)$ with a preset threshold value. If the divergence is larger than the threshold, then the image cache Φ_t will be flagged such that

$$\Delta S(\mathbb{T}, \Phi_t) > \mathsf{T}, \tag{11}$$

where T is the threshold. In this paper the threshold T is set to 0.01.

3.2   Unknown Image Discovery

The closed set oriented solutions [1, 5, 6, 26, 33] fail to address the problem of unknown image discovery because of an insufficient number of the samples shot by unknown camera models in the identification classifier training stage. In order to address this issue, this paper develops a new process named unknown image discovery. The unknown image discovery is clustering ensemble based so that it recognizes the image samples shot by unknown camera models from the image cache. Moreover, a novel parameter-based optimization algorithm is proposed to guarantee the performance of unknown image discovery.

Clustering Ensemble Based Unknown Image Discovery The basic idea of unknown image discovery is merging the images from training dataset and image cache for clustering. In the clustering solution, if a cluster does not contain any images shot by the known camera models, the images in this cluster are likely to be unknown images. This observation motivates the following unknown image discovery.

Given the training dataset $\mathbb{T}$, which consists of images from K models, $\{C_1, \cdots, C_K\}$. Suppose that Φ_t images in the image cache detected at time duration t were new unknown images. Firstly, $\mathbb{T}$ and Φ_t are merged together as $\mathbb{V}$ for clustering. The clustering solution includes k clusters: $\{c_1, \cdots, c_k\}$. Then, for a cluster c_i, if it does not contain any images from $\mathbb{T}$, then the images in the cluster c_i are labeled as unknown images. To further increase the purity of the discovered unknown samples, the strategy of clustering ensemble is adopted. An ensemble of clusters is constructed with different clustering numbers: $(k-1)$, k and $(k+1)$. Only the images identified as unknown samples in all the clustering solutions are finally recognized as unknown images. Algorithm 1 presents the procedure of unknown image discovery.

Algorithm 1: Clustering ensemble based unknown image discovery

---

Input : Training dataset : $\mathbb{T}$.
         Image cache : Φ_t.
         Optimization parameter : k.
Output: The newly discoveried unknown samples : $\Delta\mathbb{U}$.

1   $\Delta\mathbb{U} \leftarrow \emptyset$;
    // Combine the training dataset and image cache.
2   $\mathbb{V} \leftarrow \mathbb{T} \bigcup \Phi_t$;.
3   **for** $j = (k-1)$ to $(k+1)$ **do**
4 | Perform clustering on $\mathbb{V}$ to obtain clusters $\{c_1, \cdots, c_j\}$;
5 | $\mathbb{Z}_j \leftarrow \emptyset$;
6 | **for** $i = 1$ to j **do**
7 | **if** c_i does not contain any images from $\mathbb{T}$ **then**
8 | $\mathbb{Z}_j \leftarrow (\mathbb{Z}_j \bigcup c_i)$;
9 | **end**
10 | **end**
11 **end**
    // Only the images identified as samples shot by unknown camera models in all the clustering solutions are finally labeled as unknown images.
12   $\Delta\mathbb{U} \leftarrow \bigcap_{j=(k-1)}^{(k+1)} \mathbb{Z}_j$;
13 **return** $\Delta\mathbb{U}$.

More specifically, we adopt the k-means [23] algorithm for the clustering process. That is, the k-means algorithm helps us partition the set $\mathbb{V}$ into k clusters with respect to minimizing the within-cluster sum of squares. Our algorithm progresses on a set of randomly selected k centroids by switching between the two following modes: In the assignment step, our algorithm assigns each image to the cluster with the closest mean value; and in the update step, our algorithm calculates the new means of the centroid of the images in the cluster. Our algorithm will be terminated when the assignments stabilize.

Optimization of Parameter k Finding the optimal value for the key parameter is a significant challenge for statistical learning-based image source identification. In our proposed scheme, we need to optimize the parameter k that is the number of clusters used in the unknown image discovery in Algorithm 1. To find the optimal k value, we investigate the effects of changing k values. And based on the investigation, we develop an automated method to find the optimal k for unknown image discovery.

The unknown image discovery process recognizes the images that were shot by the unknown camera models. We measure its performance with True Positive Rate (TPR) and False Positive Rate (FPR). TPR is defined as the rate of the number of correctly detected unknown images over the total of images from the unknown camera models; FPR is the ratio of the number of images from the known camera models erroneously detected as unknown images to the total of images from known camera models.

Figure 4a shows the TPR and FPR of unknown image discovery with different k values. We set $s = 4000$ and $u = 3$ according to Sect. 4. To reveal the impact of different number of clusters as k values, we feed the images of three unknown camera models to the image cache. Figure 4a shows that TPR and FPR are significantly affected by different values of the parameter k. In general, both TRP and FPR increase when k increases. For example, when k increases from 100 to 1500, TPR increases from 16% to 78%; and FPR also increases from 0% to nearly 19%.

A high TPR of unknown image discovery can boost the final identification performance, but a high FPR can cause negative impact. Therefore, a good choice of k should be selected to balance TPR and FPR to achieve high identification accuracy. As shown in Fig. 4a, both TPR and FPR monotonically increase as k increases. Therefore, an optimal k is the maximum k when FPR is approximately zero.

However, FPR on Φ_t cannot be measured because the images in the cache are out of our control in practice. In this paper, we use a heuristic solution to estimate the approximate FPR. That is, we copy a small number of labeled images from the training set $\mathbb{T}$ to the image cache Φ_t and then calculate the FPR for these labeled images. We calculate the FPR in the following steps: 1) a small portion (20%) of labeled images are randomly selected from $\mathbb{T}$ as $\Delta\mathbb{T}$; 2) Φ_t and $\Delta\mathbb{T}$ are combined as $\Phi'_t = \Phi_t \bigcup \Delta\mathbb{T}$; 3) we perform the unknown image discovery on Φ'_t and use the FPR on $\Delta\mathbb{T}$ to approximate the FPR of Φ_t.

Figure 4b shows the FPR of Φ_t (denoted as FPR) and of $\Delta\mathbb{T}$ (denoted as pFPR) with different k values. The plot shows that pFPR is very close to FPR. Despite the

Fig. 4 Optimizing parameter
k for unknown image
discovery on the image cache
with 3 unknown models. (**a**)
Impact of parameter *k*. (**b**)
The FPR and pFPR. (**c**) The
optimal *k*

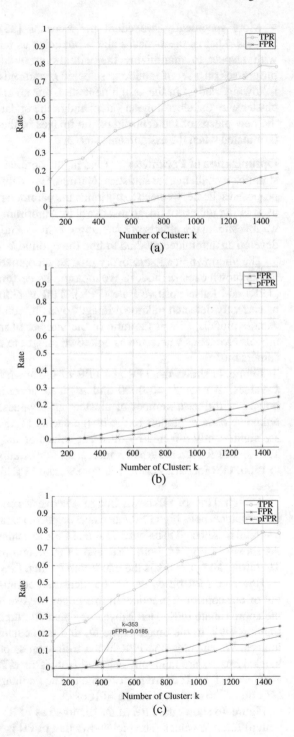

Algorithm 2: Optimizing the number of clusters k for unknown image discovery

---

Input : Training dataset : $\mathbb{T}$.
   Image Cache : Φ_t.
   The maximum number of clusters : k_{max}.
   The pFPR threshold : T_{fpr} (default 2%).
Output: The optimal k.

1   $\Delta\mathbb{T} \leftarrow$ randomly selected 20% of images from $\mathbb{T}$;
2   $\mathbb{T}' \leftarrow \mathbb{T} - \Delta\mathbb{T}$;
3   $\Phi_t' \leftarrow \Phi_t + \Delta\mathbb{T}$;
4   $k_{min} \leftarrow 1$;
5   **while** $k_{min} \neq k_{max}$ **do**
6    $k \leftarrow (k_{min} + k_{max})/2$;
    // Perform unknown image discovery using Algorithm 1 with training dataset $\mathbb{T}'$, image cache Φ_t', and parameter k.
7    $\Delta\mathbb{U}' \leftarrow$ UknownDiscovery($\mathbb{T}', \Phi_t', k$);
8    $\Delta\mathbb{X}' \leftarrow \Delta\mathbb{U}' \cap \Delta\mathbb{T}$;
9   pFPR(k) $\leftarrow \frac{|\Delta\mathbb{X}'|}{|\Delta\mathbb{T}|}$;
10   **if** pFPR(k) $< T_{fpr}$ **then**
11 | $k_{min} \leftarrow k + 1$;
12   **else if** pFPR(k) $> T_{fpr}$ **then**
13 | $k_{max} \leftarrow k - 1$;
14   **else**
15 | break;
16
17 **end**
18 $k_{opt} \leftarrow k$;
19 **return** k_{opt}.

---

fact that the actual FPR cannot be measured, pFPR can be calculated as follows:

$$\text{pFPR} = \frac{\text{\# images in } \Delta\mathbb{T} \text{ detected as unknown}}{|\Delta\mathbb{T}|}. \tag{12}$$

Considering that pFPR is slightly larger than FPR and the singular points exist in the image set, we set a threshold for pFPR, denoted by T_{fpr}. The pFPR threshold T_{fpr} is slightly larger than 0. Based on our empirical results, we find that 2% is a suitable value for T_{fpr}.

Solving an optimization problem can be time-consuming. In our case, the k value varies between 100 and 1,500. To find the optimal k quickly, we use a binary search technique that takes logarithmic time. Algorithm 2 describes the procedure of automatic parameter selection. Figure 4c shows that the optimal $k_{opt} = 353$ is automatically selected by following Algorithm 2. With this k_{opt} value, we can detect as many as unknown images while keeping the pFPR rate lower than the threshold T_{fpr}.

3.3 (K + 1)-class Classification

The existing solutions [11, 18, 21, 22, 40] detect images shot by unknown camera models in the offline scenario. For the online image source identification, the images appear continuously as an image stream with unknown images, which is not well studied in the previous work.

Different to existing solutions, OISU can detect the new unknown images via unknown sample triage and recognize the images shot by unknown camera models through the unknown image discovery process. The set of unknown images can be combined as follows:

$$\mathbb{U} = \bigcup_d \Delta \mathbb{U}_d, \tag{13}$$

where $\Delta \mathbb{U}_d$ is the unknown images discovered within time duration d.

The unknown image discovery process is subsequently incorporated in the image identification procedure by customizing a $(K + 1)$-class classification. That is, the images of the K known camera models are treated as K-class correspondingly, and the discovered unknown images are regarded as the extra 1-class. All of the testing images shot by any unknown camera models will be classified into the specific unknown class by the $(K+1)$-class classifier. Therefore, OISU identifies the images of the unknown camera models and distinguishes the images of the known camera models. In this paper, we implement the multi-class classifier by using the multi-class SVM algorithm [10, 21] because the multi-class SVM classifiers achieved good results in many existing studies [1, 5, 6, 26, 33].

4   Experiments and Results

To evaluate the performance of the proposed scheme, we carried out a large number of tests on a real-world image collection named "Dresden" [19]. This section reports our empirical studies and the results.

4.1   Dataset and Experiment Settings

The Dresden image collection [19] was used for the empirical study. We chose to use the Dresden dataset that was built for developers and researchers to benchmark camera-based digital forensic solutions. More specifically, the total 16,961 images were shot by using different brands and models of cameras in the various scenes including natural, urban, indoor, and outdoor environments. Table 1 summarizes the 27 camera models, the number of images of each model, and the aliases of the camera models.

Table 1 The Dresden image collection

No.	Model	Size	Alias	No.	Model	Size	Alias	No.	Model	Size	Alias
1	Agfa_DC – 504	169	A1	10	FujiFlm_FinePixJ50	630	F1	19	Pentax_OptioW60	192	P3
2	Agfa_DC – 733s	281	A2	11	Kodak_M1063	2391	K1	20	Praktica_DCZ5.9	1019	PR1
3	Agfa_DC – 830i	363	A3	12	Nikon_CoolPixS710	925	N1	21	Ricoh_GX100	854	R1
4	Agfa_Sensor505 – x	172	A4	13	Nikon_D200	752	N2	22	Rollei_RCP – 7325XS	589	RO1
5	Agfa_Sensor530s	372	A5	14	Nikon_D70	369	N3	23	Samsung_L74wide	686	S1
6	Canon_Ixus55	224	C1	15	Nikon_D70s	367	N4	24	Samsung_NV15	645	S2
7	Canon_Ixus70	567	C2	16	Olympus_mju_1050SW	1040	O1	25	Sony_DSC – H50	541	SO1
8	Canon_PowerShotA640	188	C3	17	Panasonic_DMC – FZ50	931	P1	26	Sony_DSC – T77	725	SO2
9	Casio_EX – Z150	925	C4	18	Pentax_OptioA40	638	P2	27	Sony_DSC – W170	405	SO3

We randomly selected 20% of all the 16,961 images as the training dataset, and the rest were treated as the testing dataset. To simulate the problem of unknown cameras, 9 models were randomly selected as the known models, that is, RO1, N3, C2, P1, K1, O1, A3, C3, and C1. Thus, the remaining 18 camera models were treated as the unknown camera models.

To simulate the scenario of online image source identification, we need to create a stream of images selected from image pools shot by both known and unknown models. We mix the images by iterating through the following steps:

- *Round 0*: testing images shot by the 9 known camera models were mixed randomly;
- *Round 1*: testing images shot by the 9 known camera models and by the $1 * u$ unknown camera models were mixed randomly;
- $\cdots$;
- *Round i*: testing images shot by the 9 known camera models and by the $i * u$ unknown camera models were mixed randomly;
- $\cdots$;
- *Last Round*: stop when the images shot by all unknown camera models were mixed,

where u is a control parameter, namely, unknown camera model increment size. We used the parameter u to control the number of newly added unknown camera models in each round. In our experiment, u varies between 2 and 4. The unknown camera models were introduced according to the following order—A5, R1, SO1, SO3, SO2, P3, N4, A1, S1, A2, A4, P2, S2, N2, C4, PR1, F1, and N1.

4.2 Features

We extracted 34 features by using the method of Kharrazi et al. [26]. The 34 features are classified into six categories—3 features in average pixel values, 3 features in RGB pair correlations, 3 features in neighbor distribution center of mass, 3 features in energy ratios of RGB pairs, 9 features in the wavelet domain statistics, and 13 features in the image quality metrics. Upon completion of the feature extraction, each image was represented by a feature vector of 34 dimensions.

Thereafter, we used the LIBSVM [7] to train SVMs with linear kernels in our experiments.

4.3 Evaluation Metrics

To measure the performance of online scenario, we used the accuracy and F-measure. Accuracy is the ratio of the number of all correctly identified images to the number of all identified images of the image cache.

$$\text{Accuracy} = \frac{\text{\# correctly identified images}}{\text{\# image cache size}}. \tag{14}$$

F-measure [20] was used to measure the identification performance for each camera model C_i, which is a combination of precision and recall,

$$\text{F} - \text{measure}^i = 2 \cdot \frac{\text{precision}^i \cdot \text{recall}^i}{\text{precision}^i + \text{recall}^i}. \tag{15}$$

More specifically, precision^i is the ratio of the number of correctly identified images from C_i to the number of the images identified from C_i,

$$\text{precision}^i = \frac{\text{\# correctly identified images from } C_i}{\text{\# images identified from } C_i}. \tag{16}$$

And recall^i is defined as the ratio of the number of correctly identified images from C_i over the total of images shot by camera model C_i,

$$\text{recall}^i = \frac{\text{\# correctly identified images from } C_i}{\text{\# images from } C_i}. \tag{17}$$

4.4 Performance of Triaging Unknown Samples

We evaluated the performance of our scheme proposed in Sect. 3.1. It aims to validate the feasibility of triaging the unknown samples in the image cache in real time. To answer this, we need to consider both detection accuracy and computation cost.

Detection Accuracy Figures 5 and 6 present the experimental results with respect to different values of unknown camera model increment size u and image cache size s. We also examined the accuracy of conventional multi-class SVM (MSVM) approach [1, 5, 6, 26, 33], the divergence of training dataset, and the image cache calculated according to Eq. 9.

Fig. 5 Unknown sample triage with a fixed image cache size s and varied unknown camera model increment size u. (**a**) $s = 4000, u = 2$. (**b**) $s = 4000, u = 3$. (**c**) $s = 4000, u = 4$

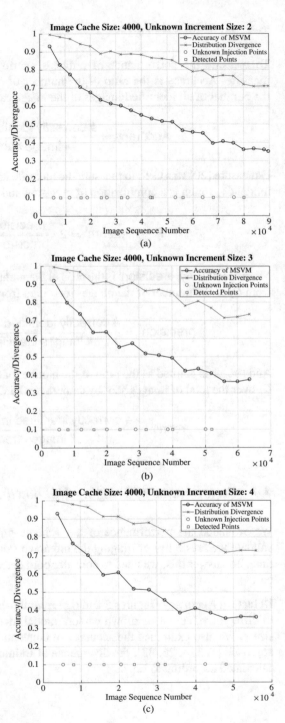

Fig. 6 Unknown sample triage with varied image cache size s and a fixed unknown camera model increment size u. (**a**) $s = 2000$, $u = 3$. (**b**) $s = 3000$, $u = 3$. (**c**) $s = 4000$, $u = 3$

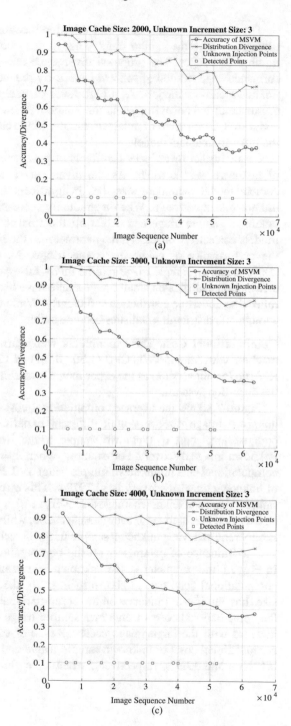

The empirical results show that the identification accuracy of MSVM and the divergence have the similar trend. That is, they both decline with the increasing number of images. The proposed divergence measurement can capture the performance degeneration of MSVM caused by the existence of images shot by any unknown camera models. At a high level of confidence, we can also detect the existence of unknown images in the image dataset. For instance, when $s = 3000$ and $u = 3$, the process of unknown sample triage can detect all the images shot by the unknown camera models.

Let us further investigate the effects of parameters s and u on the performance of unknown sample triage. As shown in Fig. 5, the parameter u, the unknown camera model increment size, has insignificant impact on the performance of unknown sample triage. When u gradually increases from 2 to 4, the process of unknown sample triage can detect all the images shot by the unknown camera models correctly. Conversely, the parameter s, the image cache size, can influence the performance of unknown sample triage. As shown in Fig. 6, when $s = 2000$, unknown sample triage returns many false positives. The reason is that the distribution statistics on small image cache have big variance and therefore cannot reflect the underline distribution sufficiently. While the size of image cache is large enough, the distribution statistics become stable.

Computational Cost The experiments were carried out on a Thinkpad T440s machine with Intel i5-4200 CPU (1.60 GHz) and 12.0 GB RAM. We used Matlab R2016b (64-bit version) as the experimental environment and Ubuntu 16.04 LTS as the operation system.

Figure 7 shows the average computational cost of unknown sample triage and unknown image discovery with different experimental parameter settings. The computational cost of unknown sample triage is significantly less than that of unknown image discovery. For example, when $s = 4000$ and $u = 3$, the average computational cost of unknown sample triage is 0.2498 s, but the computation cost of unknown image discovery is 20.2972 s. This meets the requirement of unknown sample triage and unknown image discovery: Unknown sample triage is designed to be executed in real time for every image cache, while the unknown image discovery is only required when unknown sample triage detects the new unknown samples.

The influence of parameters on the computational cost is reported in Fig. 7. In Fig. 7a, the parameter u, the unknown increment size, has little impact on the computational cost; but, as shown in Fig. 7b, the parameter s, the image cache size, has significant influence on the computational cost. As a general trend, both the computational costs of unknown sample triage and unknown image discovery increase with the large image cache sizes. For example, when $s = 2000$, the computational costs of unknown sample triage and unknown image discovery are 0.1333 s and 12.9 s, respectively; when $s = 4000$, the computational costs of

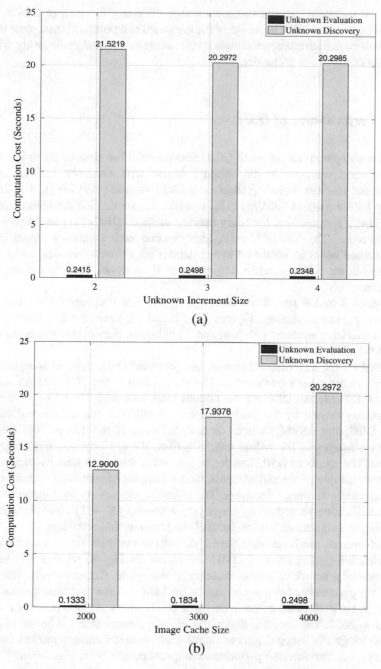

Fig. 7 Computational cost of unknown sample triage and unknown image discovery. (**a**) Varied unknown camera model increment size u and fixed image cache size $s = 4000$. (**b**) Fixed unknown camera model increment size $u = 3$ and varied image cache size s

unknown sample triage and unknown image discovery increase to 0.2498 s and 20.2972 s, respectively. The reason of the increased computation is that more images are involved in both unknown sample triage and unknown image discovery with the increase of the image cache size.

4.5 Performance of OSIU

We also compared our proposed OSIU scheme with the state-of-art image source identification solutions in the online scenario with unknown camera models. Traditional schemes including multi-class SVM method (MSVM) [1, 5, 6, 26, 33], binary SVMs method (BSVM) [18], combined classification framework method (CCF) [40], and decision boundary carving method (DBC) [11] are implemented as references. The Source Camera Identification with Unknown model (SCIU) [22] method needs an unlabeled image dataset for unknown mining, which is not available in the online scenario. Therefore, SCIU does not work well in the online situation.

Figures 8 and 9 report the accuracy of the five comparative methods with different parameter settings. Figures 10, 11, and 12 report the F-measures of the five comparative methods with 3, 6, and 12 unknown camera models, when we set $s = 4000$ and $u = 3$.

Based on the experimental results, our proposed OSIU scheme is significantly superior to the other four methods. The second best is the CCF method. And the MSVM, BSVM, and DBC are the bottom three methods. Their low performance was mainly caused by the inadequate consideration of the unknown class. The CCF, DBC, and BSVM methods do not make use of the dynamically emerging unknown images in the online scenario; thus, the performance improvement is limited. The proposed OSIU can detect the new unknown images dynamically via unknown sample triage and recognize some samples of unknown camera models through unknown image discovery. The unknown images are then incorporated in the identification procedure by employing a special $(K + 1)$-class classification, resulting in a significant improvement of the identification accuracy.

Furthermore, we investigate the impact of the two parameters s and u on the identification performance of OSIU. As shown in Fig. 8, when s is fixed, the influence of u on identification accuracy is negligible. Comparatively, when u is fixed, the parameter s has a big impact on the identification performance. As shown in Fig. 9, the performance improvement of OSIU with $s = 4000$ is larger than that with $s = 2000$. The reason is that when the image cache size s is larger, unknown sample triage can recognize more samples of unknown camera models from the image cache; therefore, the performance improvement is more significant.

Fig. 8 Accuracy of compared methods with a fixed image cache size s and varied unknown increment size u. (**a**) $s = 4000$, $u = 2$. (**b**) $s = 4000$, $u = 3$. (**c**) $s = 4000$, $u = 4$

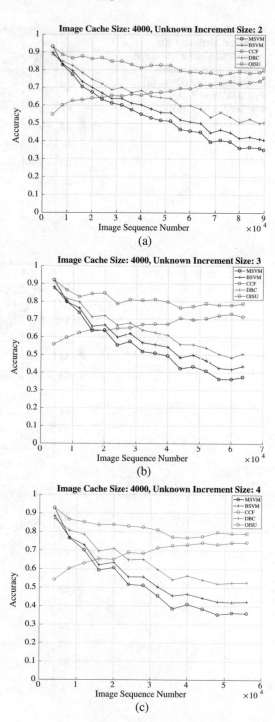

Y. Huang et al.

Fig. 9 Accuracy of
compared methods with
varied image cache size s and
a fixed unknown camera
model increment size u. (**a**)
$s = 2000$, $u = 3$. (**b**)
$s = 3000$, $u = 3$. (**c**)
$s = 4000$, $u = 3$

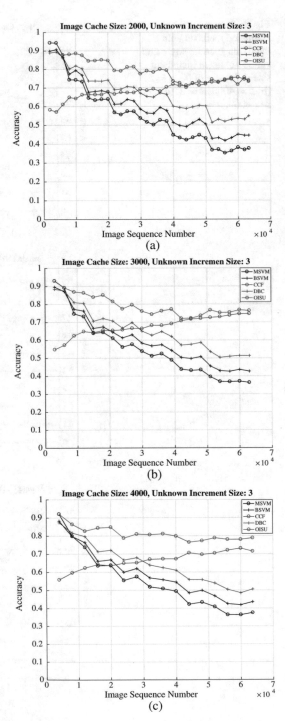

(a)

(b)

(c)

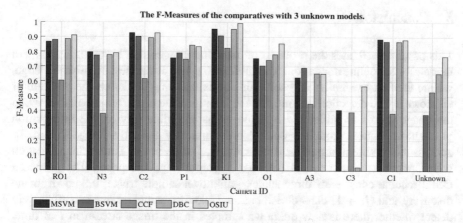

Fig. 10 The F-measure of the compared solutions with 3 unknown camera models

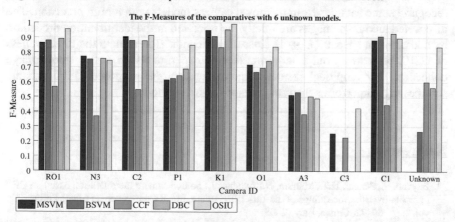

Fig. 11 The F-measure of the compared solutions with 6 unknown camera models

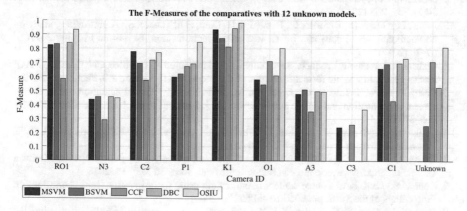

Fig. 12 The F-measure of the compared solutions with 12 unknown camera models

5   Conclusion

This paper investigates the problem of detecting images shot by unknown camera models for the online image source identification task. To the best of our knowledge, we are the first to identify and solve this issue. Existing works on the problem of unknown image detection assume that the identification is performed offline. When the image source identification is deployed as an online service, the information of unknown camera models is emerging dynamically along the image stream. To address this problem, a new scheme, namely, Online image Source Identification with Unknown camera model (OSIU), is proposed in this paper. The proposed OSIU scheme consists of three stages—unknown sample triage, unknown image discovery, and $(K + 1)$-class classification. Unknown sample triage can effectively detect whether there is new unknown samples in the image stream in real time. When any new unknown sample is detected, unknown image discovery can recognize some samples from unknown camera models with a high precision. The discovered unknown images are finally incorporated in the identification procedure by addressing a special $(K + 1)$-class classification. To evaluate the effectiveness of OSIU, we carried out a large number of experiments on a real-world image collection. Our empirical results demonstrate that the proposed OSIU scheme significantly outperforms the four state-of-the-art methods.

References

1. Bayram, S., Sencar, H.T., Memon, N., Avcibas, I.: Source camera identification based on CFA interpolation. In: Proceedings of the 12th IEEE International Conference on Image Processing, vol. 3, pp. 69–72. Genoa, Italy (2005)
2. Bondi, L., Baroffio, L., Güera, D., Bestagini, P., Delp, E.J., Tubaro, S.: First steps toward camera model identification with convolutional neural networks. IEEE Signal Process. Lett. **24**(3), 259–263 (2017)
3. Bondi, L., Güera, D., Baroffio, L., Bestagini, P., Delp, E.J., Tubaro, S.: A preliminary study on convolutional neural networks for camera model identification. Electronic Imaging **2017**(7), 67–76 (2017)
4. Camera and Imaging Products Association Standardization Committee and Others (CIPA): Exchangeable image file format for digital still cameras: EXIF Version 2.3, Tech. Rep, 2010
5. Castiglione, A., Cattaneo, G., Cembalo, M., Petrillo, U.F.: Experimentations with source camera identification and online social networks. J. Ambient Intell. Humaniz. Comput. **4**(2), 265–274 (2013)
6. Çeliktutan, O., Sankur, B., Avcibaş, I.: Blind identification of source cell-phone model. IEEE Trans. Inf. Forensics Secur. **3**(3), 553–566 (2008)
7. Chang, C.C., Lin, C.J.: LIBSVM: a library for support vector machines, National Taiwan University, http://www.csie.ntu.edu.tw/~cjlin/libsvm/, accessed May 15, 2018
8. Choi, K.S., Lam, E.Y.: Source camera identification using footprints from lens aberration. In: Proceedings of the SPIE, pp. 60690J1–60690J8 (2006)
9. Choi, K.S., Lam, E.Y., Wong, K.K.Y.: Automatic source camera identification using the intrinsic lens radial distortion. Optics Express **14**(24), 11551–11565 (2006)
10. Cortes, C., Vapnik, V.: Support-vector networks. Machine Learning **20**(3), 273–297 (1995)

11. Costa, F.D., Silva, E., Eckmann, M., Scheirer, W.J., Rocha, A.: Open set source camera attribution and device linking. Pattern Recogn. Lett. **39**, 92–101 (2014)
12. Cox, J.I., Miller, M.L., Bloom, J.A.: Digital Watermarking. Morgan Kaufmann Publishers Inc., San Francisco, CA, USA (2002)
13. Dirik, A.E., Sencar, H.T.: Source camera identification based on sensor dust characteristics. In: Proceedings of the IEEE Workshop on Signal Processing Applications for Public Security and Forensics, pp. 1–6. Washington, DC, USA (2007)
14. Dirik, A.E., Sencar, H.T., Memon, N.: Digital single lens reflex camera identification from traces of sensor dust. IEEE Trans. Inf. Forensics Secur. **3**(3), 539–552 (2008)
15. Freire-Obregón, D., Narducci, F., Barra, S., Castrillón-Santana, M.: Deep learning for source camera identification on mobile devices. Preprint (2017). arXiv:1710.01257
16. Gan, G., Ma, C., Wu, J.: Data Clustering: Theory, Algorithms, and Applications. SIAM (2007)
17. Geradts, Z.J., Bijhold, J., Kieft, M., Kurosawa, K., Kuroki, K., Saitoh, N.: Methods for identification of images acquired with digital cameras. In: Proceedings of the Enabling Technologies for Law Enforcement, pp. 505–512 (2001)
18. Gloe, T.: Feature-based forensic camera model identification. In: Transactions on Data Hiding and Multimedia Security, vol. VIII, pp. 42–62. Springer, Berlin, Heidelberg (2012)
19. Gloe, T., Böhme, R.: The Dresden image database for benchmarking digital image forensics. J. Digit. Forensic Pract. **3**(2–4), 150–159 (2010)
20. Hripcsak, G., Rothschild, A.S.: Agreement, the F-measure, and reliability in information retrieval. J. Am. Med. Inform. Assoc. **12**(3), 296–298 (2005)
21. Hsu, C.W., Lin, C.J.: A comparison of methods for multiclass support vector machines. IEEE Trans. Neural Netw. **13**(2), 415–425 (2002)
22. Huang, Y., Zhang, J., Huang, H.: Camera model identification with unknown models. IEEE Trans. Inf. Forensics Secur. **10**(12), 2692–2704 (2015)
23. Jain, A.K.: Data clustering: 50 years beyond K-means. Pattern Recogn. Lett. **31**(8), 651–666 (2010)
24. Kang, X., Chen, J., Lin, K., Peng, A.: A context-adaptive SPN predictor for trustworthy source camera identification. EURASIP J. Image Video Process. **2014**(1), 1–11 (2014)
25. Kang, X., Li, Y., Qu, Z., Huang, J.: Enhancing source camera identification performance with a camera reference phase sensor pattern noise. IEEE Trans. Inf. Forensics Secur. **7**(2), 393–402 (2012)
26. Kharrazi, M., Sencar, H.T., Memon, N.: Blind source camera identification. In: Proceedings of the 11th IEEE International Conference on Image Processing, pp. 709–712. Singapore (2004)
27. Kullback, S., Leibler, R.A.: On information and sufficiency. Ann. Math. Stat. **22**(1), 79–86 (1951)
28. Li, C.T.: Source camera identification using enhanced sensor pattern noise. IEEE Trans. Inf. Forensics Secur. **5**(2), 280–287 (2010)
29. Li, C.T., Li, Y.: Color-decoupled photo response non-uniformity for digital image forensics. IEEE Trans. Circuits Syst. Video Technol. **22**(2), 260–271 (2012)
30. Lukáš, J., Fridrich, J., Goljan, M.: Digital camera identification from sensor pattern noise. IEEE Trans. Inf. Forensics Secur. **1**(2), 205–214 (2006)
31. Pan, L., Trepanic, N.: Image source detection: A case study on Facebook images taken by iPhones. In: Proceedings of the Second Applications and Techniques in Information Security Workshop (ATIS'11), pp. 39–46
32. Park, M.G.H.H.J., Har, D.H.: Source camera identification based on interpolation via lens distortion correction. Aust. J. Forensic Sci. **46**(1), 98–110 (2014)
33. Peng, F., Shi, J., Long, M.: Comparison and analysis of the performance of PRNU extraction methods in source camera identification. J. Comput. Inf. Syst. **9**(14), 5585–5592 (2013)
34. Stamm, M.C., Liu, K.J.R.: Forensic detection of image manipulation using statistical intrinsic fingerprints. IEEE Trans. Inf. Forensics Secur. **5**(3), 492–506 (2010)
35. Sutcu, Y., Bayram, S., Sencar, H.T., Memon, N.: Improvements on sensor noise based source camera identification. In: Proceedings of the IEEE International Conference on Multimedia and Expo, pp. 24–27. Beijing, China (2007)

36. Tsai, M.J., Wang, C.S., Liu, J., Yin, J.S.: Using decision fusion of feature selection in digital forensics for camera source model identification. Comput. Stand. Interfaces **34**(3), 292–304 (2012)
37. Tuama, A., Comby, F., Chaumont, M.: Camera model identification with the use of deep convolutional neural networks. In: Proceedings of the 2016 IEEE International Workshop on Information Forensics and Security, pp. 1–6. Abu Dhabi, United Arab Emirates (2016)
38. Van, L.T., Emmanuel, S., Kankanhalli, N.S.: Identifying source cell phone using chromatic aberration. In: The IEEE International Conference on Multimedia and Expo, pp. 883–886. Beijing, China (2007)
39. van Schyndel, R.G., Tirkel, A.Z., Osborne, C.F.: A digital watermark. In: Proceedings of the 1st IEEE International Conference Image Processing, vol. 2, pp. 86–90. Austin, Texas, USA (1994)
40. Wang, B., Kong, X., You, X.: Source camera identification using support vector machines. In: Advances in Digital Forensics, vol. V, pp. 107–118. Springer, Berlin, Heidelberg (2009)
41. Wu, T.F., Lin, C.J., Weng, R.C.: Probability estimates for multi-class classification by pairwise coupling. J. Mach. Learn. Res. **5**(Aug), 975–1005 (2004)

Reinforcement Learning Based Communication Security for Unmanned Aerial Vehicles

Liang Xiao, Donghua Jiang, and Sicong Liu

1 Introduction

With the rapid development of unmanned aerial vehicles (UAVs) in communications, networking, and sensing applications, UAVs have gained considerable research interest in the last decade. Although UAV applications have been widely applied in many different fields, especially the military surveillance and environment monitoring, UAV communication process is not sufficiently safe due to jamming attacks. By imposing jamming signals on the controller during the communication process of the drones, a jammer can interfere with the sensing data reception of the controller, exhaust the drone battery, or keep the drone from following the specified sensing mission waypoint [18].

The state-of-the-art technologies such as frequency hopping [22] and smart antenna [2] have been employed to resist jamming attacks. For example, in a zero-sum game of drone pursuit and escaping, the Isaacs' method is utilized to determine the drone trajectory against jamming attacks as proposed in [1]. The anti-jamming scheme of drone power control in [32] formulates a Bayesian Stackelberg game and uses a sub-gradient method to develop an iterative algorithm and determine the drone transmit power. Nevertheless, these drone anti-jamming schemes require to be aware of the jamming and channel models, and the performance is affected by the versatility and accuracy of the knowledge of these models on the drones.

In order to cope with the fact that the drone has little knowledge about the channel model, some experts use reinforcement learning (RL) to optimize the anti-jamming

Part of this work was supported by NSFC No. 61971366 and 61901403.

L. Xiao (✉) · D. Jiang · S. Liu
Department of Communication Engineering, Xiamen University, Xiamen, Fujian, China
e-mail: lxiao@xmu.edu.cn; liusc@xmu.edu.cn

57

strategy through repeated anti-jamming tests in the drone communication process. For example, the UAV anti-jamming scheme named QPC in [12] uses the Q-learning algorithm to determine the drone transmit power and resist jamming attacks. This proposed scheme can improve the quality of the received signals and meanwhile reduce the energy consumption of the drone in the presence of reactive jamming attacks. Nevertheless, this solution does not consider the situation where the drone sensing mission requires a waypoint, nor does it optimize the drone trajectory in the transmission for drone sensing applications.

One part of this chapter is focused on applying the deep RL algorithms to address the issue of jamming attacks, and an RL-based drone trajectory and power control (RLTPC) scheme is proposed for the drone to jointly choose its trajectory and transmit power, optimize the system performance, and meet the requirements of the sensing mission waypoint. In this scheme, the drone determines the transmit strategy (including the trajectory and transmit power) in the sensing mission according to its current position, the next position in the waypoint, and the communication feedback from the controller. In the case when the channel between the jammer and the controller and the jamming model are unknown at the drone, a multilayer perceptron (MLP) is applied to obtain an action-value function that approximates the system state. This scheme evaluates the quality of satisfaction (QoS) of the UAV sensing mission based on the signal-to-interference-plus-noise ratio (SINR) sent back by the receiver. Moreover, the drone utilizes the evaluated utility to determine the next communication strategy against jamming attacks. The drone sensing mission for a given waypoint and channel model according to [24] are investigated by simulations. Compared with the benchmark of QPC in [12], the proposed trajectory and power control scheme can save the drone energy consumption and significantly improve the QoS of the drone communication against intelligent jamming attacks.

Another part of this book is focused on applying the prospect theory (PT) to study smart attacks on drone transmission initiated by selfish and subjective attackers. In order to reveal the mechanism of how the utility of the drone is related with the subjectivity of Eve, we have derived the Nash equilibria (NE) of the PT-based game of smart attacks and the conditions for the existence of the NE. Moreover, because the repeated game can be expressed as a Markov decision process (MDP), RL algorithms can be used to derive the optimal strategy to resist smart attacks without being aware of the network model. More specifically, we utilize the Q-learning algorithm in the power allocation strategy, in which the reward of each state-action pair in the learning process is evaluated by the Q-function. At the same time, the power allocation scheme based on the algorithm of Win or Learn Faster-Policy Hill Climbing (WoLF-PHC) is proposed to evaluate the uncertainty of drone transmission against smart attacks and improve the safety of the drones. There are a lot of attack states, feasible transmit power levels, and channel conditions in the system of drone transmission in the presence of jamming attacks, which requires a faster learning speed compared with the Q-leaning or WoLF-PHC-based solutions. A power allocation scheme based on the deep Q-network (DQN) as investigated in [15] accelerates the learning rate of the WoLF-PHC-based solution. Moreover, deep learning and Q-learning algorithms are combined to compress the state space to cope with high dimensional issues in the learning process. The simulation results show that the proposed anti-jamming scheme using the DQN algorithm improves the SINR of the received signal, secrecy capacity, and utility of the drone.

2 Communication Security for Unmanned Aerial Vehicles

2.1 UAV Communication Model

Case 1 There is a drone communication network in a three-dimensional space that consists of a legitimate drone, a controller, and a smart jammer. The drone in the network has access to M waypoints denoted by $\{W_m\}_{1 \leq m \leq M}$, which is devised for performing some tasks like point checking. The waypoints perform tasks sequentially at the scheduled times denoted by $\{nT\}$ with $1 \leq n \leq M$ and T is a fixed time interval between two tasks. The target waypoints of the drone are the same as the flight between two adjacent waypoints. During the flight, the drone tends to arrange its trajectory closer to the controller located at G to improve the communication quality and reach the next waypoint on time. In a time slot, the maximum velocity of the drone in the x, y, and z dimensions is represented by v. When the drone arrives at the waypoint as planned, it detects the situation of the environment and transmits the sensing data to the controller [9, 18].

Case 2 In the UAV communication network, a transmission model between the legal drone and the mobile ground unit (MGU) against smart attacks is constructed. The drone is intended to transmit monitoring data and messages to the MGU. The distance between the MGU and the UAV is d_t. The number of the radio frequency channels occupied is B. Meanwhile, the smart attacker equipped with a radio device is able to perform smart attacks and selects an attack pattern.

The power gain of the i-th transmission channel between the drone and the MGU at time k is $h_{T,i}^{(k)}$, the power gain of the i-th wiretap channel is $h_{E,i}^{(k)}$, and the power gain of the i-th jamming channel is $h_{J,i}^{(k)}$. Thus the channel power gain between the drone and the MGU at time k is formulated as $\mathbf{H}_T^{(k)} = [h_{T,i}^{(k)}]_{1 \leq i \leq B}$, and the wiretap channel gain from the drone to the target drone is $\mathbf{H}_E^{(k)} = [h_{E,i}^{(k)}]_{1 \leq i \leq B}$. Similarly, the jamming channel gain from the jammer to the MGU is $\mathbf{H}_J^{(k)} = [h_{J,i}^{(k)}]_{1 \leq i \leq B}$.

According to [17], the reference distance of the drone transmission is denoted by d_0 with ξ being a reference path loss at d_0, and the path loss exponent of the channel model is denoted by ρ. The path loss of the transmission channel denoted by PL with the transmitter–receiver distance d can be given by

$$PL(\text{dB}) = \xi(\text{dB}) + 10\rho \lg\left(\frac{d}{d_0}\right), \quad d > d_0. \tag{1}$$

2.2   Attack Model

Smart jammer observes the ongoing UAV communication to estimate the current UAV transmission states and decides whether to send faked signals with power $p_J^{(k)}$ to the target ground node at time k or not. The channel gain from the smart jammer to the ground node is $h_J^{(k)}$, and the maximum jamming power is P_J.

A smart attacker with programmable radio devices selects the attack mode $\mathbf{y}^{(k)}$ including eavesdropping, spoofing, and jamming attacks, etc. The attacker cannot launch eavesdropping and jamming attacks at the same time because jamming signals are sent with the same frequency with the target UAV, which prevents the attacker from eavesdropping the ongoing transmit information. Smart attackers can compromise MGUs to jam or spoof downlink UAV communication, or launch jamming attacks to interrupt the uplink UAV communication. For example, the compromised UAV can move close to the target UAV before sending faked signals to decrease the detection accuracy. The compromised MGU can move close to the target UAV to improve the jamming performance. The downlink UAV sensing information transmission is proposed as an example, while this scheme can be applied to other UAV communication schemes as well.

The attack mode over B frequency channels is $\mathbf{y}^{(k)} = [y_i^{(k)}]_{1 \leq i \leq B}$. More specifically, the compromised MGU transmits faked signals with the transmit power $y_i^{(k)} \in [0, \cdots, P_J]$ on frequency channel i, if $\mathbf{y}^{(k)} > 0$; a smart attacker eavesdrops the target UAV if $\mathbf{y}^{(k)} = -1$; the smart attacker conducts spoofing attack to Bob if $\mathbf{y}^{(k)} = -2$; otherwise, the smart attacker does not apply attacks on the ongoing transmission.

3   Reinforcement Learning Based UAV Communication Security

3.1   Reinforcement Learning Based Anti-Jamming Communications

As an MDP, the UAV anti-jamming communication process in terms of the trajectory control and power control can apply reinforcement learning to improve its utility based on the energy consumption and communication quality. In the RL-based UAV anti-jamming communication scheme, the action policy of the UAV is determined according to the state at each time slot, which consists of the position, target waypoint, and previous transmission quality of the UAV obtained from the SINR in the feedback information. The state is then input to a deep Q-network to infer the long-term discounted reward, which is called Q-value, of all feasible UAV polices in this state. The UAV will then select a transmit policy based on the Q-values, update the weights of the deep Q-network according to the utility, and observe the next state.

At each time slot k, the current target waypoint of the UAV is denoted by $w^{(k)} \in \mathcal{W}$. The UAV obtains its current position information $l^{(k)}$, which is denoted by a three-dimensional coordinate, according to the Global Position System. From the feedback of the ground node, the UAV can obtain the SINR of the last received signal, which is denoted by $\rho^{(k-1)}$. The state of the UAV includes its position, the target waypoint, and the SINR of last received signal, i.e., $s^{(k)} = \left\{ l^{(k)}, w^{(k)}, \rho^{(k-1)} \right\}$. Let $\mathcal{S}$ denote the state space that consists of all the possible states.

As illustrated in Fig. 1, at each time slot k, the UAV inputs the state to a convolutional neural network (CNN) known as a deep Q-network [15], which is parameterized by θ, to estimate the Q-value of any legal action x at the current state. The action includes the movement vector and the transmit power, i.e., $x^k = \left[\left(v^{(k)}\right)^T, p_U^{(k)} \right]^T$. As the deep Q-network cannot handle continuous actions, the three dimensions of the position and the transmit power are quantized into N_x, N_y, N_z, and N_p levels, respectively. Let $\mathcal{X}$ denote the action space that consists of all the feasible actions.

The action with the maximum Q-value is the optimal choice at the current state. However, with such a greedy strategy of action selection, the UAV is much likely to fall into local minima due to the lack of exploration of the global state space and action space. Thus, in practical implementation, the UAV usually applies the ε-greedy scheme for action selection based on the Q-values $Q\left(x | s^{(k)}, \theta^{(k)}\right), \forall x \in \mathcal{X}$ at current states $s^{(k)}$. Specifically, the UAV chooses the action with the maximum Q-value with a large probability $1 - \varepsilon$ and keeps a small probability ε to randomly select another action in the action space to guarantee a thorough exploration.

Algorithm 1: RL-based anti-jamming trajectory and power control scheme

---

1: Initialize $s^{(0)}$ $\gamma, \alpha, C_1, C_2, \theta, \hat{\theta} = \theta, \rho^{(0)}$ and $\mathcal{R} = \emptyset$
2: **for** $k = 1, 2, \cdots$ **do**
3:   Obtain position of the UAV
4:   Obtain the SINR of the last received signal
5:   Formulate the current state $s^{(k)} = \left\{ l^{(k)}, w^{(k)}, \rho^{(k-1)} \right\}$
6:   Input current state to the Q-network to estimate $Q\left(s^{(k)}, x; \theta\right), \forall x \in \mathcal{X}$
7:   Determine the trajectory and power via the ε-greedy policy
8:   Conduct a flight according to $v^{(k)}$ and sense the environment data
9:   Send the environment data to the ground node at $p_U^{(k)}$
10: Calculate utility via Eq. (2)
11: Append the experience to memory pool
12: Sample experiences at random from the memory pool
13: Update the network weights via Eq. (3)
14: **end for**

The UAV flies toward the next position according to $v^{(k)}$ and sends the environment data to the receiver at the power $p_U^{(k)}$ during the flight. Then the UAV evaluates QoS by comparing the SINR with a predefined threshold σ and calculates

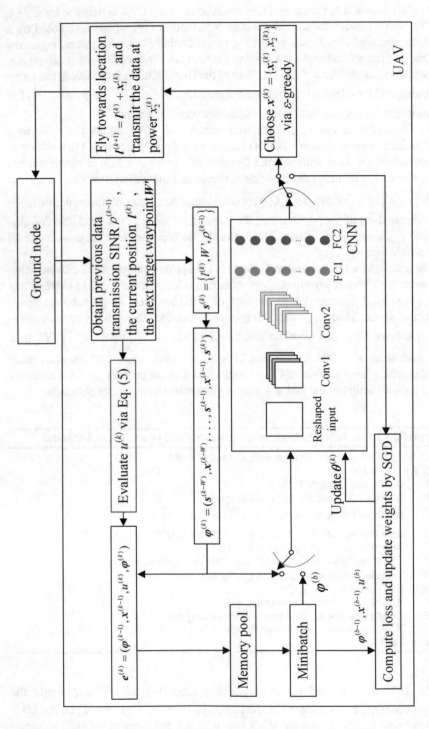

Fig. 1 RLTPC scheme for the UAV against smart jamming attack

the energy consumption for the flight and transmission. The UAV can also measure the distance to $w^{(k)}$ with the equipped position sensors. Then, the UAV can calculate the utility as

$$u^{(k)} = I\left(\rho^{(k-1)} - \sigma\right) - \delta\left(\left\|l^{(k)} - w^{(k)}\right\|\right) - C_1\left\|v^{(k)}\right\| - C_2 p_U^{(k)}, \qquad (2)$$

where δ, C_1, and C_2 are the positive coefficients used for balancing the cost of energy consumption and punishment for the distance. The function $I(\cdot)$ is an indicator function that equals to 1 if the condition therein is true and 0 otherwise.

The state, action, utility, and the next state can be summarized as a piece of experience at each time slot, which is $e^{(k)} = \left\{s^k, x^{(k)}, u^{(k)}, s^{k+1}\right\}$. The UAV stores the experience in a memory pool denoted by $\mathcal{R}$. At each time slot, the UAV randomly samples N experiences from the memory pool to update the current network weights $\theta^{(k)}$. Specifically, the UAV first estimates the target Q-values according to the Bellman equation based on the utilities and the Q-values of the next time slot from a target Q-network, whose weight is denoted by $\hat{\theta}^{(k)}$, and calculates the loss function according to the mean square error between the Q-values and the target Q-values. Then it updates $\theta^{(k)}$ using the gradient descent algorithm with the loss function, which can be given by

$$\theta^{(k)} \leftarrow \theta^{(k)} + \alpha \nabla_\theta \mathbb{E}\left[\left(u^{(i)} + \gamma \max_{x \in \mathcal{X}} Q\left(s^{(i+1)}, x; \hat{\theta}^{(k)}\right)\right.\right.$$

$$\left.\left. - Q\left(s^{(i)}, x^{(i)}; \theta^{(k)}\right)\right)^2\right], \qquad (3)$$

where $\alpha \in (0, 1)$ controls the learning speed of the Q-network and $\gamma \in (0, 1)$ is the discount factor. Every τ steps, the UAV copies the weights of the Q-network and updates the target network [15]. The proposed RLTPC scheme is summarized in Algorithm 1.

The performance of the proposed RLTPC scheme is assessed via simulations in a specific UAV communication network against a smart jammer as shown in Fig. 2. For simplicity, a concrete wireless channel model is adopted to evaluate the channel gain and a classic jamming policy is conducted. Besides, RLTPC can be applied to other environments with different channels and jammers as well.

As illustrated in Fig. 2, the initial position of the UAV is $(0, 50, 30)$ m in a cuboid space whose length, width, and height are 200 m, 100 m, and 50 m, respectively. The UAV uses sensors including a thermometer and a camera, etc. to collect the environment data. There are four waypoints for the UAV to inspect located in $(40, 50, 30)$ m, $(80, 50, 30)$ m, $(120, 50, 30)$ m, and $(160, 50, 30)$ m in sequence. The UAV must transmit the sensing data to the ground node located at the origin and move from one waypoint to the next in 4 seconds. The available transmit power ranges from 0 to 4 W. The UAV determines its trajectory and transmit power in the fight every 1 second according to the RLTPC scheme. The UAV can fly along the

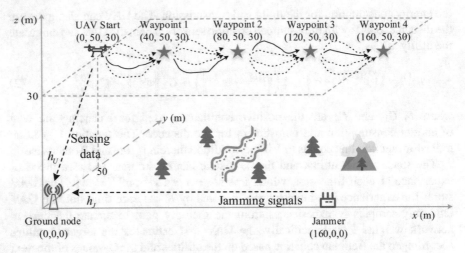

Fig. 2 Illustration of the simulation settings of the anti-jamming UAV communication network

three directions or the opposite directions for 0, 10, or 20 m at each time slot. In the utility function, the coefficients C_1 and C_2 are 0.25 and 0.1, respectively. The threshold σ is set as 15 dB to evaluate the QoS. In the RLTPC scheme, the learning rate is 10^{-4} and the discounted factor is 0.99. The UAV samples 32 experiences from the memory pool, whose size is 10000, at each time slot.

Assuming that the UAV works in a hilly environment, the channel model is formulated according to [24]. The carrier frequency of the transmission signal is 5.06 GHz, and thus the path loss exponent is 1.7 in the channel model. Besides, since the distance between the ground node and the UAV is less than the reference distance of 3.4 km, the channel has a path loss of 119.7 dB with a large-scale fading X, which follows a Gaussian distribution whose variance is 2.4 dB. Thus, the instant channel gain can be written as

$$h_L = 10^{-\frac{119.7+X}{10}} \left(\frac{\|l\|_2}{3400} \right)^{-1.7}. \tag{4}$$

Ricean fading channel is applied to model the small-scale fading, which is denoted by h_S whose K-factor is calculated according to [24]. The channel gain includes both h_L and h_S.

In the simulations, the jammer is located at (160, 0, 0) m, and it can sense the ongoing signals and transmit noisy signals immediately. The jamming power p_J is chosen from 5 levels including 20, 40, 60, 80, and 100 mW. The jammer calculates its utility with respect to each available jamming power according to the QoS of the signal from the UAV to the ground node and its energy consumption, which can be written as

$$u_a = -\mathrm{I}\left(\rho^{(k)} > \sigma \right) - 10p_J. \tag{5}$$

Then the jammer chooses the jamming power that maximizes its utility.

Reinforcement learning has been applied to a similar environment in [12] using a QPC scheme. In the QPC scheme, the state includes the jamming power and channel gain, and the action to be determined is the transmit power of the UAV. The trajectory of the UAV is predetermined and will not change dynamically with the jammer. It is obvious that the UAV can further improve the transmission quality and reduce the energy consumption by jointly adjusting the flight direction and the transmit power in real time.

As illustrated in Fig. 3, the RLTPC scheme can improve the QoS and reduce the energy consumption. As shown in Fig. 3a and b, the UAV using the RLTPC scheme increases the QoS by 9.5% and reduces the energy consumption by 42.3% compared with the benchmark of QPC in [12] at the 30000-th time slot. As shown in Fig. 3c, the utility of the UAV with the proposed scheme finally converges to 0.663, while that with the QPC scheme is 0.328, which is about 50.5% lower. Owing to the deep Q-network and the experience replay technique, the convergence time of the RLTPC scheme is 16.7% less than that of QPC, even though both the state space and action space of the RLTPC scheme are more complex.

3.2 Reinforcement Learning Based UAV Communications Against Smart Attacks

In this section, in order to obtain a user-centric view of smart attacks, we formulate a dynamic smart attack game based on the PT for UAV communications. Meanwhile, the drone needs to determine the power distribution strategy to resist the smart attacks, so it adopts RL techniques to choose the optimal strategy to maximize the system reward. By using RL techniques, the drone with the identification Alice can make her decision without being aware of the communication channel model. In the following three sections, we will introduce three different RL-based methods to determine the power allocation strategies, i.e., Q-learning, WoLF-PHC, and DQN algorithms.

Q-Learning Based Power Allocation In this part, Alice uses the Q-learning algorithm to determine the power allocation strategies. Alice selects the transmit power denoted by $\mathbf{x}$ according to the network state denoted by $\mathbf{s}$ and the transmission history in each time slot and then sends signals to the MGU called Bob. For simplicity, the transmit power value is quantified into $10P_T + 1$ levels, where P_T is the total power constraint of Alice. At time slot k, Alice observes the system state $\mathbf{s}^{(k)}$ composed of the previous attack mode and then chooses the optimal action $\mathbf{x}^{(k)}$ according to the Q-function $Q\left(\mathbf{s}^{(k)}, \mathbf{x}^{(k)}\right)$. Alice chooses the action according to the ϵ-greedy strategy, where ϵ is a very small value between 0 and 1 to balance the exploration and the exploitation. More specifically, Alice chooses the power value to maximize its $Q\left(\mathbf{s}^{(k)}, \mathbf{x}^{(k)}\right)$ with the high probability $1 - \epsilon$. Meanwhile, Alice determines the power value at random with ϵ.

Fig. 3 Performance of the
UAV with the RLTPC scheme
compared with the
benchmark. (**a**) Quality of
satisfaction. (**b**) Energy
consumption for flight and
data transmission. (**c**) Utility
of the UAV

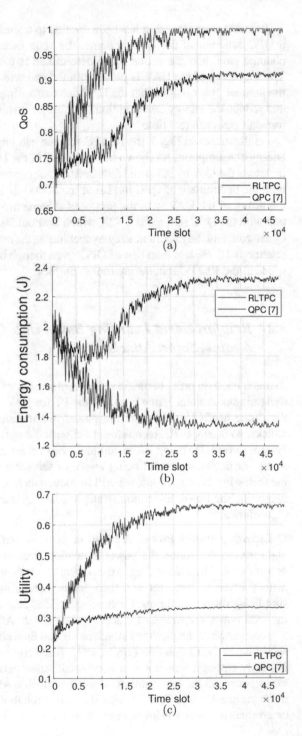

After transmitting the signal to Bob, Alice gets feedback from the environment and needs to update the Q-function. As a typical model-free RL algorithm, the Q-learning algorithm regards the Q-function as the expected long-term discount utility in terms of the transmit power $\mathbf{x}$ and the state $\mathbf{s}$. Based on the iterative Bellman equation, Alice updates $Q\left(\mathbf{s}^{(k)}, \mathbf{x}^{(k)}\right)$ as given by

$$Q\left(\mathbf{s}^{(k)}, \mathbf{x}^{(k)}\right) \leftarrow (1 - \tau) Q\left(\mathbf{s}^{(k)}, \mathbf{x}^{(k)}\right)$$

$$+ \tau \left(u \left(\mathbf{s}^{(k)}, \mathbf{x}^{(k)}\right) + \gamma \max_{\mathbf{x}^{(k)} \in \mathbf{X}} Q \left(\mathbf{s}^{(k+1)}, \mathbf{x}^{(k+1)}\right) \right), \quad (6)$$

where τ is the learning factor between 0 and 1 representing the efficiency of the learning, and γ is the discount factor between 0 and 1 measuring the importance of the future rewards.

WoLF-PHC-Based Power Allocation In this part, Alice uses the WoLF-PHC-based power allocation strategy to determine the transmit power allocation that confuses Eve at random. Different from the Q-learning, this algorithm extends to a hybrid strategy game and utilizes fast win or learning principles and learning parameters to achieve better convergence [3]. More specifically, Alice uses a mixed-strategy table $\pi(\mathbf{s}^{(k)}, \mathbf{x})$ to determine the transmit strategy. The sum of the elements in the table is 1.

In order to update the table, Alice uses two learning parameters denoted by δ_w and $\delta_l \in (0, 1]$ for different cases. If Alice loses, she must learn faster and understand Eve's policy more quickly. That is, the speed of learning δ_l must be faster than the speed of winning δ_w. The number of state is $C(\mathbf{s}^{(k)})$. Then the average hybrid strategy table $\bar{\pi}$ is given by

$$\bar{\pi}\left(\mathbf{s}^{(k)}, \mathbf{x}\right) \leftarrow \bar{\pi}\left(\mathbf{s}^{(k)}, \mathbf{x}\right) + \frac{1}{C(\mathbf{s}^{(k)})} \left(\pi\left(\mathbf{s}^{(k)}, \mathbf{x}\right) - \bar{\pi}\left(s^{(k)}, \mathbf{x}\right)\right), \forall \mathbf{x} \in \mathbf{X}. \quad (7)$$

If Alice obtains a higher value than that of the average hybrid strategy, she will win and update the table with the rate δ_w; otherwise, Alice will lose and update the blending strategy table with the other rate δ_l. Thus we have

$$\delta^{(k)} = \begin{cases} \delta_w, & \sum_{\mathbf{x} \in \mathbf{X}} \pi\left(\mathbf{s}^{(k)}, \mathbf{x}\right) Q\left(\mathbf{s}^{(k)}, \mathbf{x}\right) > \sum_{\mathbf{x} \in \mathbf{X}} \bar{\pi}\left(\mathbf{s}^{(k)}, \mathbf{x}\right) Q\left(\mathbf{s}^{(k)}, \mathbf{x}\right) \\ \delta_l, & \text{o.w..} \end{cases} \quad (8)$$

Then the system needs to update the hybrid strategy table π

$$\pi\left(\mathbf{s}^{(k)}, \mathbf{x}\right) \leftarrow \pi\left(\mathbf{s}^{(k)}, \mathbf{x}\right)+$$

$$\begin{cases} -\min\left(\pi\left(\mathbf{s}^{(k)}, \mathbf{x}\right), \frac{\delta^{(k)}}{|\mathbf{X}|-1}\right), & \text{if } \mathbf{x} \neq \underset{\hat{\mathbf{x}} \in \mathbf{X}}{\arg\max}\, Q\left(\mathbf{s}^{(k)}, \hat{\mathbf{x}}\right) \\ \sum_{\mathbf{x} \neq \hat{\mathbf{x}}} \min\left(\pi\left(\mathbf{s}^{(k)}, \mathbf{x}\right), \frac{\delta^{(k)}}{|\mathbf{X}|-1}\right), & \text{o.w.,} \end{cases} \tag{9}$$

where $\delta^{(k)}$ represents the learning parameter.

Based on the above hybrid strategy table, Alice chooses $\mathbf{x}^{(k)}$ according to

$$\Pr\left(\mathbf{x}^{(k)} = \hat{\mathbf{x}}\right) = \pi\left(\mathbf{s}^{(k)}, \hat{\mathbf{x}}\right), \quad \forall \hat{\mathbf{x}} \in \mathbf{X}. \tag{10}$$

After transmitting with the chosen power, Alice can receive Eve's response consisting of the BER or packet loss rate measured by Bob and then infer both the SINR value and the attack mode. Thus Alice uses the obtained data to calculate its utility and update the Q-function via (6).

DQN-Based Power Allocation The above strategies need to evaluate the Q-function, and the required learning time will increase rapidly as the state space increases. The size of the state space depends on B, P_T, and P_J, which will all increase with the increase of the quantization precision requirement. To accelerate the learning process, we propose a DQN-based power allocation scheme using CNNs.

As shown in Fig. 4, the CNN of the DQN-based strategy is composed of two convolutional layers with rectified linear units, followed by two fully connected

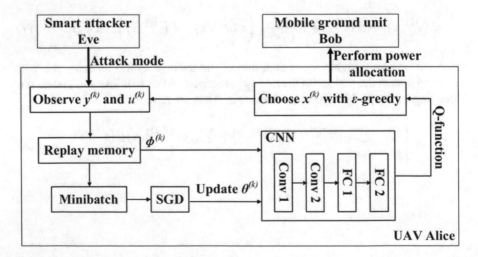

Fig. 4 DQN-based drone power allocation scheme

layers. We set W as the time window. More specifically, in the first $W-1$ time slots, Alice chooses the strategy among the action space at random. Afterwards, the drone chooses the power allocation strategy based on the ϵ-greedy policy. Thus, the drone chooses the optimal action based on a DQN to estimate the Q-function of each policy.

After transmitting the signals with the selected strategy, Alice observes the attack mode and obtains the system utility. Then, Alice needs to update the DQN with the experience pool named replay memory. A state sequence is composed of the current system state and the previous W state-action pairs at time k, denoted by $\varphi^{(k)}$. Then the experience composed of the state sequence $\varphi^{(k)}$, the chosen action $\mathbf{x}^{(k)}$, the utility $u^{(k)}$, and the next state sequence $\varphi^{(k+1)}$ is formulated as $e^{(k)}$, which is put into a reply buffer $\mathcal{D} = \{e^{(j)}\}_{1 \leq j \leq k}$. Instead of using the most recent experiences, Alice uses random minibatches in the data set to update the network. More specifically, an experience denoted by $e^{(d)}$ is randomly chosen from the reply buffer $\mathcal{D}$ for T times to update the network weights. According to the stochastic gradient descent (SGD) algorithm, the network weights denoted by $\theta^{(k)}$ can be updated by minimizing the loss function as

$$\theta^{(k)} = \arg\min_{\theta} \mathbb{E}_{\varphi,\mathbf{x},u,\varphi'}\left[\left(u^{(k)} + \gamma \max_{\mathbf{x}'} Q(\varphi',\mathbf{x}';\theta^{(k-1)}) - Q(\varphi,\mathbf{x};\theta)\right)^2\right].$$

(11)

Simulation Results To verify the performance of the power allocation strategy proposed above, several groups of practical simulation experiments are carried out. In the simulation experiments, Eve uses a Q-learning based strategy to conduct attack, such as spoofing, launch jamming, and eavesdropping attacks. For typical cases, the system parameters are configured according to [5], and thus, we set $P_T = P_J = 0.4, \sigma = 1, C_m = -0.5, [\beta_l]_{0 \leq l \leq 5} = [0.1, 0.8, 0.05, 0.03, 0.02, 0]$, $[\eta_l]_{0 \leq l \leq 5} = [0.1\ 0.6\ 0.1\ 0.05\ 0.05\ 0.1], \alpha_E = 0.8, \alpha_A = 1, \tau = 0.95, \gamma = 0.7$, $d_0 = 10$ m, and $\epsilon = 0.9$. As for the transmission parameters, we select $d_e = 30$ m, $d_t = d_j = 50$ m, $\xi = 0.02, 0.075$, and 0.0032 for the Alice–Bob, Eve–Bob, and Alice–Eve transmission links, respectively. Due to the approximate free-space propagation, both the Alice–Bob and Alice–Eve links have $\rho = 2$. According to a two-ray model, the Eve–Bob channel has $\rho = 4$.

As shown in Fig. 5, the channel gains randomly change every 300 time slots in the first experiment. The safe rate of the system, which is the possibility of being free from attacks, increases over time, which means that the system is becoming safer if the proposed strategies are performed. For instance, after 1500 time slots, the safe rates of the three strategies are increased by about 25%. Obviously, compared with the WoLF-PHC and Q-learning based strategies, the DQN-based one has the highest safe rate. More specifically, the safe rate of the DQN-based one is 93%, 7% and 11% higher than those based on the WoLF-PHC and Q-learning, respectively. Meanwhile, the secrecy capacity of the DQN-based one is higher than that of the WoLF-PHC and Q-learning strategies. Furthermore, the SINR of the DQN-based strategy is higher than the SINR of the WoLF-PHC and Q-learning based

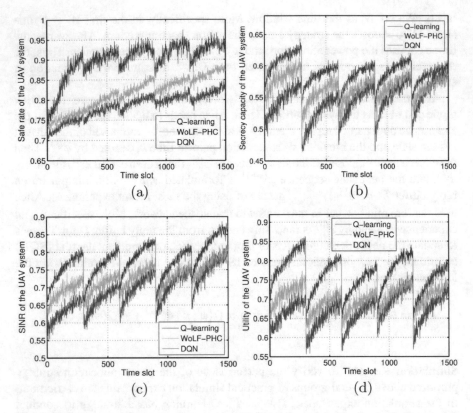

Fig. 5 Performance of the proposed strategies with $P_T = P_J = 0.4$, $L = 5$, $\sigma = 1$, $C_m = -0.5$, $\alpha_E = 0.8$, and $\alpha_A = 1$. (**a**) Safe rate. (**b**) Secrecy capacity. (**c**) SINR. (**d**) Utility

strategies. If the channel gain changes over time, the utility will first decrease and then increase rapidly. The utility of the DQN-based one increases by 0.84 compared to the beginning value, which is about 22% and 13% higher than that of Q-learning and WoLF-PHC. Overall, as shown in Fig. 6, the DQN-based algorithm reduces the impact of the spoofing attack to the best extent over time compared to Q-learning and WoLF-PHC strategies.

In the second experiment as shown in Fig. 7, we try to verify from Eve's subjective perspective of view that the proposed scheme can improve the UAV communication quality. As shown by the results, the average safe rate over many times of experiments is reduced by 16.3% compared with the Q-learning algorithm if α_E changes from 0.6 to 1. The DQN-based one outperforms the WoLF-PHC and Q-learning strategies with a higher safe rate. For example, the average safe rate of the DQN-based one is 11% and 18% higher than the WoLF-PHC and Q-learning based strategies if $\alpha_E = 0.9$. Meanwhile, the average secrecy capacity decreases with α_E. For instance, as for the DQN-based one, the average secrecy capacity decreases by 5.3% as α_E changes from 0.9 to 1. Moreover, the average

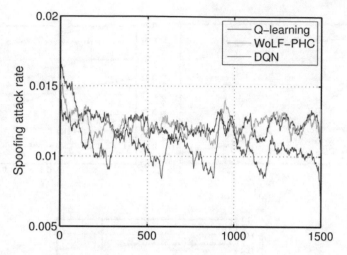

Fig. 6 The impact of the smart attack in the drone system with $P_T = P_J = 0.4$, $L = 5$, $\sigma = 1$, $C_m = -0.5$, $\alpha_E = 0.8$, and $\alpha_A = 1$

SINR decreases with α_E. For example, if we set $\alpha_E = 0.9$, the DQN-based strategy can increase the value by 10.3% compared to the Q-learning one.

We perform the third experiment to show the impact of the total power constraint of Eve as shown in Fig. 8 with $P_T = 0.4$. The average safe rate decreases with the attacker's total power. For example, the average safe rate of the Q-learning based strategy is reduced by 20.5% if P_J changes from 0.3 to 0.7 because there is more power for the attacker to deteriorate the legitimate drone communications. The DQN-based strategy has the highest average SINR, which is because the average SINR of the DQN-based strategy are 22.3% and 29.3% higher than the WoLF-PHC and Q-learning based ones, respectively, if $P_J = 0.7$.

4 UAV Secure Communication Game

4.1 Game Model

The interaction between the target UAV and the smart attacker can be formulated as a UAV transmission game $\mathcal{G}$, in which the smart attacker selects an attack mode and the target UAV chooses the transmit power x_i. The UAV controller implemented with physical layer security mechanisms [14] can detect spoofing attacks conducted by the smart attacker, and the miss detection rate is β. The jamming detection method based on the packet delivery ratio and the received signal strength has a miss detection rate η for the jamming attacks [13]. In addition, the target UAV can hardly detect eavesdropping attacks.

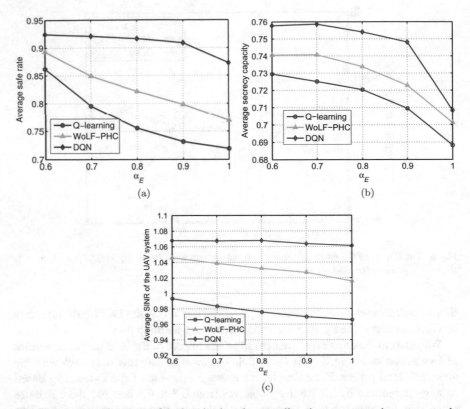

Fig. 7 Average performance of the learning based power allocation strategy against smart attacks with $P_T = P_J = 0.4$, $L = 5$, $\sigma = 1$, $C_m = 1.5$, and $\alpha_A = 1$. (**a**) Average safe rate. (**b**) Average secrecy capacity. (**c**) Average SINR of the UAV system

The importance of the UAV transmission energy in this UAV communication scheme is reflected by the energy consumption factor μ. The UAV energy consumption C_m is based on the miss detection rate of spoofing attacks. The noise power is σ. The target UAV aims to improve the ongoing transmission quality, such as the SINR of the UAV signals, and increase its utility $u_A^{(k)}$. More specifically, if $\mathbf{y} = -\mathbf{1}$, we have

$$u_A^{(k)}(\mathbf{x}, \mathbf{y}) = \sum_{i=1}^{B} \left(h_{T,i}^{(k)} - h_{E,i}^{(k)} - \mu \right) x_i, \tag{12}$$

where the first right term is the UAV communication secrecy capacity. If $\mathbf{y} = -\mathbf{2}$,

$$u_A^{(k)}(\mathbf{x}, \mathbf{y}) = \sum_{i=1}^{B} \left(h_{T,i}^{(k)} - \mu \right) x_i - \beta C_m. \tag{13}$$

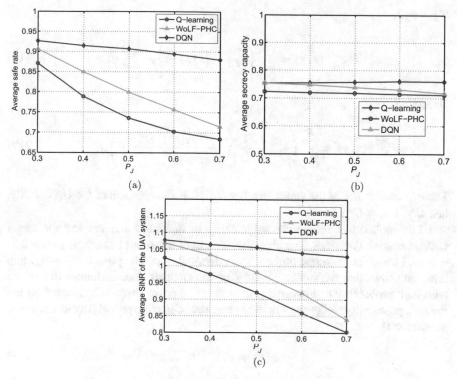

Fig. 8 Average performance of the learning based power allocation strategy against smart attacks with $P_T = 0.4$, $L = 5$, $\sigma = 1$, $C_m = -0.5$, $\alpha_E = 0.8$, and $\alpha_A = 1$. (a) Average safe rate. (b) Average secrecy capacity. (c) Average SINR of the UAV system

Otherwise,

$$u_A^{(k)}(\mathbf{x}, \mathbf{y}) = \sum_{i=1}^{B} \left((1 - \eta) h_{T,i}^{(k)} + \frac{\eta h_{T,i}^{(k)}}{\sigma + h_{J,i}^{(k)} y_i} - \mu \right) x_i. \tag{14}$$

More specifically, the UAV detection accuracy is quantized into $L + 1$ levels. The UAV has the detection accuracy $p = l/L$ with the probability $\beta_l = \Pr(\beta = p)$ for spoofing attacks and $\eta_l = \Pr(\eta = p)$ for jamming attacks. In addition, $\beta_l(\eta_l) > 0$ and $\sum_{l=0}^{L} \beta_l(\eta_l) = 1$. If $\mathbf{y} = -\mathbf{1}$, the EUT-based UAV utility is given by

$$U_A^{EUT}(\mathbf{x}, \mathbf{y}) = \sum_{i=1}^{B} \left(h_{T,i}^{(k)} - h_{E,i}^{(k)} - \mu \right) x_i. \tag{15}$$

If $\mathbf{y} = -2$,

$$U_A^{EUT}(\mathbf{x}, \mathbf{y}) = \sum_{i=1}^{B} \left(h_{T,i}^{(k)} - \mu \right) x_i - \frac{C_m}{L} \sum_{l=1}^{L} l\beta_l. \tag{16}$$

Otherwise,

$$U_A^{EUT}(\mathbf{x}, \mathbf{y}) = \sum_{i=1}^{B} \left(h_{T,i}^{(k)} - \frac{1}{L} \sum_{l=1}^{L} \frac{l\eta_l h_{T,i}^{(k)} h_{J,i}^{(k)} y_i}{\sigma + h_{J,i}^{(k)} y_i} - \mu \right) x_i. \tag{17}$$

The expected utility of the smart attacker U_E^{EUT} is the opposite of the UAV utility, i.e., $U_E^{EUT} = -U_A^{EUT}$.

PT is applied to consider the subjectivity of the smart attacker and the target UAV in making decisions. The objective weight of the smart attacker is denoted as $\alpha \in (0, 1]$, and the objective probability is denoted as p. The probability weighting function shows that the subjective UAV or the attacker underestimates the events with high probability and overstates the events with low probability. Based on the Prelec's probability weighting function [16], the subjective probability of the smart attacker w is

$$w(p) = e^{-(-\ln p)^\alpha}. \tag{18}$$

Smart attackers tend to use a subjective perspective of view to choose the attack mode in uncertain security transmission games, and the PT-based UAV utility U_A^{PT} replaces the objective probability in the EUT-based utility U_A^{EUT} with the subjective probability given by

$$U_A^{PT}(\mathbf{x}, \mathbf{y}) = \sum_{i=1}^{B} \left(h_{T,i}^{(k)} - h_{E,i}^{(k)} - \mu \right) x_i, \tag{19}$$

if $\mathbf{y} = -1$.

$$U_A^{PT}(\mathbf{x}, \mathbf{y}) = \sum_{i=1}^{B} \left(h_{T,i}^{(k)} - \mu \right) x_i - \frac{C_m}{L} \sum_{l=1}^{L} l w_A(\beta_l), \tag{20}$$

if $\mathbf{y} = -2$. Otherwise,

$$U_A^{PT}(\mathbf{x}, \mathbf{y}) = \sum_{i=1}^{B} \left(h_{T,i}^{(k)} - \frac{1}{L} \sum_{l=1}^{L} l w_A(\eta_l) \frac{h_{T,i}^{(k)} h_{J,i}^{(k)} y_i}{\sigma + h_{J,i}^{(k)} y_i} - \mu \right) x_i. \tag{21}$$

The PT-based utility of the smart attacker U_E^{PT} is given by

$$U_E^{PT}(\mathbf{x}, \mathbf{y}) = \sum_{i=1}^{B} \left(h_{E,i}^{(k)} - h_{T,i}^{(k)} + \mu \right) x_i, \tag{22}$$

if $\mathbf{y} = -1$.

$$U_E^{PT}(\mathbf{x}, \mathbf{y}) = \sum_{i=1}^{B} \left(\mu - h_{T,i}^{(k)} \right) x_i + \frac{C_m}{L} \sum_{l=1}^{L} l w_E(\beta_l), \tag{23}$$

if $\mathbf{y} = -2$. Otherwise,

$$U_E^{PT}(\mathbf{x}, \mathbf{y}) = \sum_{i=1}^{B} \left(\frac{1}{L} \sum_{l=1}^{L} l w_E(\eta_l) \frac{h_{T,i}^{(k)} h_{J,i}^{(k)} y_i}{\sigma + h_{J,i}^{(k)} y_i} + \mu - h_{T,i}^{(k)} \right) x_i. \tag{24}$$

The index k is omitted in the rest of book for simplicity without ambiguity.

4.2 Nash Equilibrium of the Game

The smart attacker with a subjective decision strategy chooses the transmission strategy to make a trade-off between the risk of being detected and the damage to the UAV system. Therefore, the smart attacker is aimed at maximizing the PT-based utility instead of the EUT-based one. Let $\mathbf{I}_{(a,b)}$ denote an a-dimensional all-zero vector except being one at the b-th element.

Theorem 1 *The PT-based UAV communication game $\mathcal{G}$ has an NE $(P_T \mathbf{I}_{(B,j^*)}, -1)$ if*

$$\max_{1 \le i \le B} \{ h_{T,i} - h_{E,i} \} > \mu, \tag{25}$$

$$L h_{E,j^*} P_T > \max \left\{ C_m \sum_{l=0}^{L} l w_E(\beta_l), \frac{h_{T,j^*} h_{J,j^*} P_T P_J}{\sigma + h_{J,j^*} P_J} \sum_{l=0}^{L} l w_E(\eta_l) \right\}, \tag{26}$$

where

$$j^* = \arg \max_{1 \le v \le B} (h_{T,v} - h_{E,v}). \tag{27}$$

Proof By (19), if (25) holds, we have

$$U_A^{PT}\left(P_T\mathbf{I}_{(B,j^*)}, -\mathbf{1}\right) = P_T \max_{1 \le i \le B} \{h_{T,i} - h_{E,i} - \mu\}$$

$$\ge \sum_{i=1}^{B}(h_{T,i} - h_{E,i} - \mu)x_i = U_A^{PT}(\mathbf{x}, -\mathbf{1}). \tag{28}$$

If (26) holds, we have

$$U_E^{PT}\left(P_T\mathbf{I}_{(B,j^*)}, -\mathbf{1}\right) = -P_T\left(h_{T,j^*} - h_{E,j^*} - \mu\right)$$

$$> \max\left\{(\mu - h_{T,j^*})P_T + \frac{C_m}{L}\sum_{l=0}^{L}lw_E(\beta_l),\right.$$

$$\left.(\mu - h_{T,j^*})P_T + \frac{1}{L}\frac{h_{T,j^*}h_{J,j^*}P_TP_J}{\sigma + h_{J,j^*}P_J}\sum_{l=0}^{L}lw_E(\eta_l)\right\}$$

$$= \max\left\{U_E^{PT}\left(P_T\mathbf{I}_{(B,j^*)}, -\mathbf{2}\right), U_E^{PT}\left(P_T\mathbf{I}_{(B,j^*)}, \mathbf{y} \ge \mathbf{0}\right)\right\}. \tag{29}$$

Therefore, $(P_T\mathbf{I}_{(B,j^*)}, -\mathbf{1})$ is an NE of the PT-based UAV communication game $\mathcal{G}$.

Corollary 1 *If (25) and (26) hold, the utility of the UAV in the PT-based UAV communication game $\mathcal{G}$ is*

$$U_A^{EUT} = P_T \max_{1 \le i \le B}\{h_{T,i} - h_{E,i}\} - \mu P_T. \tag{30}$$

Remark The smart attacker eavesdrops the ongoing communications and avoids being detected if the wiretap channel has a good condition. If another channel has a better condition than the wiretap one, i.e., $\max_i\{h_{T,i} - h_{E,i}\} > \mu$, the UAV transmits the sensing information with the maximum transmit power on the channel, which has the maximum gap of channel gain between the transmission and the wiretap channels, to improve the UAV communication secrecy capacity.

Theorem 2 *The PT-based UAV communication game $\mathcal{G}$ has an NE $(P_T\mathbf{I}_{(B,q^*)}, -\mathbf{2})$ if*

$$\max_{1 \le i \le B}\{h_{T,i}\} > \mu, \tag{31}$$

$$C_m\sum_{l=0}^{L}lw_E(\beta_l) > \max\left\{Lh_{E,q^*}P_T, \frac{h_{T,q^*}h_{J,q^*}P_TP_J}{\sigma + h_{J,q^*}P_J}\sum_{l=0}^{L}lw_E(\eta_l)\right\}, \tag{32}$$

where

$$q^* = \arg\max_{1 \le v \le B} h_{T,v}. \tag{33}$$

Proof By (20), if (31) holds, we have

$$U_A^{PT}(P_T \mathbf{I}_{(B,q^*)}, -2) = P_T(h_{T,q^*} - \mu) - \frac{C_m}{L} \sum_{l=0}^{L} l w_A(\beta_l)$$

$$\ge \sum_{i=1}^{B} (h_{T,i} - \mu) x_i - \frac{C_m}{L} \sum_{l=0}^{L} l w_A(\beta_l) = U_A^{PT}(\mathbf{x}, -2). \tag{34}$$

If (32) holds, we have

$$U_E^{PT}(P_T \mathbf{I}_{(B,q^*)}, -2) = (\mu - h_{T,q^*}) P_T + \frac{C_m}{L} \sum_{l=0}^{L} l w_E(\beta_l)$$

$$> \max \left\{ (\mu - h_{T,q^*}) P_T + \frac{1}{L} \frac{h_{T,q^*} h_{J,q^*} P_T P_J}{\sigma + h_{J,q^*} P_J} \sum_{l=0}^{L} l w_E(\eta_l), \right.$$

$$\left. (\mu + h_{E,q^*} - h_{T,q^*}) P_T \right\}$$

$$= \max \left\{ U_E^{PT}(P_T \mathbf{I}_{(B,q^*)}, \mathbf{y} \succeq \mathbf{0}, U_E^{PT}(P_T \mathbf{I}_{(B,q^*)}, -1) \right\}. \tag{35}$$

Therefore, $(P_T \mathbf{I}_{(B,q^*)}, -2)$ is an NE of the PT-based UAV communication game $\mathcal{G}$.

Corollary 2 *If (31) and (32) hold, the utility of the UAV in the PT-based UAV communication game $\mathcal{G}$ is*

$$U_A^{EUT} = P_T \max_{1 \le i \le B} \{h_{T,i}\} - \mu P_T - \frac{C_m}{L} \sum_{l=0}^{L} l \beta_l. \tag{36}$$

Remark If the spoofing cost is exaggerated and the miss detection rate is large, smart attacker spoofs UAVs. The UAV allocates all the transmit power on the frequency channel with best condition to transmit sensing information.

Theorem 3 *The PT-based UAV communication game $\mathcal{G}$ with one frequency channel has an NE (P_T, P_J), if*

$$\frac{h_T h_J P_J}{\sigma + h_J P_J} \sum_{l=0}^{L} l w_E(\eta_l) > \max\left\{ L h_E, \frac{C_m}{P_T} \sum_{l=0}^{L} l w_E(\beta_l) \right\}, \tag{37}$$

$$\frac{h_T h_J P_J}{\sigma + h_J P_J} \sum_{l=0}^{L} l w_A(\eta_l) < L(h_T - \mu). \tag{38}$$

Proof By (22) to (24) and (37), as $B = 1$, we have

$$U_E^{PT}(P_T, P_J) = (\mu - h_T)P_T + \frac{1}{L} \frac{h_T h_J P_T P_J}{\sigma + h_J P_J} \sum_{l=0}^{L} l w_E(\eta_l)$$

$$> \max\left\{ (\mu + h_E - h_T) P_T, (\mu - h_T)P_T + \frac{C_m}{L} \sum_{l=0}^{L} l w_E(\beta_l) \right\}$$

$$= \max\left\{ U_E^{PT}(P_T, -1), U_E^{PT}(P_T, -2) \right\}. \tag{39}$$

If (38) holds, we have

$$\frac{\partial U_A^{PT}(x, y)}{\partial x} = h_T - \mu - \frac{1}{L} \frac{h_T h_J y}{\sigma + h_J y} \sum_{l=0}^{L} l w_A(\eta_l) > 0. \tag{40}$$

Therefore, (P_T, P_J) is an NE of the PT-based UAV communication game $\mathcal{G}$.

Corollary 3 *If $B = 1$, and (37) and (38) hold, the utility of the UAV in the PT-based UAV communication game $\mathcal{G}$ is*

$$U_A^{EUT} = (h_T - \mu) P_T - \frac{1}{L} \frac{h_T h_J P_T P_J}{\sigma + h_J P_J} \sum_{l=0}^{L} l \eta_l. \tag{41}$$

Remark The smart attacker is likely to attack the UAV transmission with a high probability if the miss detection rate is high. The UAV transmits signals with the maximum transmit power if the detection accuracy is considered to be high.

As shown in Fig. 9, the expected utility of the UAV increases with P_T and decreases at $\alpha_E = 0.792$ sharply with $P_T = 0.5$, since the smart attacker chooses eavesdropping rather than spoofing. If the transmit power is high, the smart attacker decides to spoof UAV.

Fig. 9 Performance of the PT-based UAV communication game $\mathcal{G}$ with 3 frequency channels, energy consumption 1.5, energy consumption factor 0.2, noise power 1, miss detection rate of spoofing attack [0.1, 0.8, 0.05, 0.03, 0.02, 0], miss detection rate of jamming attack [0.1, 0.6, 0.1, 0.05, 0.05, 0.1], and objective weight $\alpha = 1$

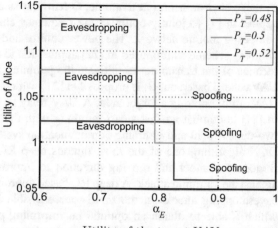

Utility of the target UAV

5 Related Work

5.1 General Anti-jamming Policies in UAV-Aided Communication

Power control is a traditional method for anti-jamming communication in wireless communication networks. The UAV power control algorithm as proposed in [32] chooses the optimal transmit power based on the sub-gradient based Bayesian Stackelberg iterative algorithm at each fixed location to address the issue of jamming attacks. Besides, due to the high mobility, the UAV can take advantage of the spatial diversity to escape from the jamming attack. The trajectory planning algorithm as designed in [1] formulates the UAV motion model against the jammer and obtains the saddle point scheme based on the Isaacs' approach. The UAVs equipped with smart antennas can further mitigate the jamming attack through specific communication methods. For example, the UAV can use the adaptive beam nulling scheme to eliminate jamming attacks, apply the Kalman filter to estimate the location of the jammer, and keep the noise signals in the null region according to [2].

5.2 Reinforcement Learning in Anti-jamming Communication

Reinforcement learning has been widely used in wireless communication networks to resist jamming attacks because of its capability to converge to the optimal policy including power control, radio frequency selection, and trajectory planning without accurate channel models or jamming models that are difficult to estimate in a highly dynamic environment. For example, Q-learning as a widely used RL algorithm is applied in [7] to jointly optimize the frequency channels for the transmitter in a competing mobile network. The power control and location selection based anti-jamming scheme as proposed in [12] applies RL to decide the transmit power and location of the transmitter to mitigate the jamming attack. A framework of secure UAV control based on RL is proposed in [21], which is able to prevent unauthorized UAVs from entering a target area. A UAV relay method for VANETs is proposed in [31] to combat against smart jamming with the help of the RL techniques. A two-dimensional anti-jamming communication system proposed in [8] applies the DQN algorithm, one of the most famous deep RL algorithms, to determine the frequency channel and moving direction to improve the communication quality against the jamming attack. A deep RL-based approach is proposed in [28] to avoid eavesdropping attacks in the multi-source visible light communications system, which is able to obtain an optimal beamforming policy through learning. Some specific techniques can be used in RL to accelerate the learning process. The Dyna-Q and Hotbooting-Q algorithms are used in [26] to determine the power for multiple antennas in a NOMA communication system and achieve a faster convergence rate and better performance compared with traditional Q-learning schemes. RL can also be applied in agents with limited computation resources such as UAVs. A deep RL-based anti-jamming method for UAV-aided cellular communications is proposed in [11] to help cellular systems resist smart jamming without being aware of the jamming model and the network model.

Multiple nodes need to take actions cooperatively to better mitigate the jamming attack in the wireless communication networks, where the multi-agent RL algorithms can be utilized. For example, the cognitive femtocell system in [19] uses a multi-agent Q-learning algorithm to determine the transmit power of each node, which achieves a faster convergence speed and a larger aggregated capacity than the system with traditional independent learning algorithm. Another femtocell network system as presented in [20] applies Q-learning to jointly determine the resource allocation and power control strategies, taking advantage of the current strategy of the neighboring nodes to improve the learning speed and system capacity. The Q-learning based power control scheme in [34] reaches a balance between the energy consumption and system performance in a multicore processor by dynamically adjusting the idle period of each core.

5.3   Game Theory in Anti-jamming Communication

The interaction among the legitimate transmitter, the receiver, and the illegal jammer is usually formulated as a game and analyzed via the game theory. For example, game theory has been used to optimize the distributed radio resource allocation in full-duplex communication systems [23] and cognitive radio networks [4, 25]. The mobile offloading game among a mobile transmitter, a security agent, and a smart attacker based on the expected utility theory is formulated in [29], in which the transmitter needs to determine its offloading rate, the security agent needs to determine its security mechanism, and meanwhile the attacker can launch jamming attacks or spoofing attacks. The interaction between a mobile user and an attacker that will launch eavesdropping or jamming attacks is formulated as a noncooperative game in [36]. The NE of the game is also analyzed via a fictitious play-based algorithm. A Bayesian game is modeled with unknown eavesdropping capacity in [6], in which the adversary has to choose which user to eavesdrop upon due to its restricted eavesdropping capacity, and a base station allocates power to users to maximize the total secrecy capacity.

The cognitive radio networks as presented in [33] utilize the prospect theory to derive the subjective price and channel partition of each user. The prospect theory is also used in [27] and [30] to analyze the influence of the subjectivity of the jammer and the end users in the cognitive radio networks with uncertain channel gains and UAV transmission, respectively. The prospect theory is used for data pricing in [10] and discloses the fact that the end-user deviation from the expected utility theory will reduce the throughput of the network. A spectrum investment game is analyzed in [35] via the prospect theory, and a conclusion is drawn that the subjective secondary operator is more likely to obtain a smaller gain with a smaller risk in the game.

6 Conclusion

In this chapter, we have investigated an RL-based power allocation and trajectory scheme for UAV transmission against jamming attacks. More specifically, the UAV applies the RL-based scheme to determine the next trajectory and select the transmit power of the sensing information without being aware of the channel and the smart jamming attack models. Simulation results show that the proposed UAV transmission scheme is able to improve the QoS of the sensing information and reduce the UAV energy consumption. For example, the proposed scheme improves the QoS in the UAV transmission by 9.5% and reduces the energy consumption by 42.3% compared with the benchmark.

The PT-based UAV transmission game considering the subjectivity of the smart attacker has been investigated in this chapter. The NE of the static game has revealed the influence of the subjectivity on the anti-jamming transmission. We have developed a deep RL-based power control scheme for the target UAV to resist smart attacks and further reduce the learning time. Simulation results show that the proposed scheme improves the secrecy capacity of the UAV system against smart attackers. For instance, the proposed scheme increases the secrecy capacity by 16% compared with the benchmark.

References

1. Bhattacharya, S., Başar, T.: Game-theoretic analysis of an aerial jamming attack on a UAV communication network. In: Proc. Amer. Control Conf., pp. 818–823. Baltimore, MD (June 2010)
2. Bhunia, S., Sengupta, S.: Distributed adaptive beam nulling to mitigate jamming in 3D UAV mesh networks. In: Proc. IEEE Int. Conf. Computing Networking Commun. (ICNC), pp. 120–125. Santa Clara, CA (January 2017)
3. Bowling, M., Veloso, M.: Multiagent learning using a variable learning rate. Artificial Intelligence 136(2), 215–250 (April 2002)
4. El-Bardan, R., Sharma, V., Varshney, P.K.: Game theoretic learning for optimal power allocation in the presence of a jammer. In: IEEE Global Conf. Signal and Info. Processing. Washington, DC (December 2016)
5. Garnaev, A., Trappe, W.: The eavesdropping and jamming dilemma in multi-channel communications. In: Proc. IEEE Int'l Conf. Commun. (ICC), pp. 2160–2164. Budapest (Jun 2013)
6. Garnaev, A., Trappe, W.: Secret communication when the eavesdropper might be an active adversary. In: Int'l Workshop on Multiple Access Commun., pp. 121–136. Springer (Aug 2014)
7. Gwon, Y., Dastangoo, S., Fossa, C., Kung, H.: Competing mobile network game: Embracing antijamming and jamming strategies with reinforcement learning. In: Proc. IEEE Conf. Commun. Network Security (CNS), pp. 28–36. National Harbor, MD (October 2013)
8. Han, G., Xiao, L., Poor, H.V.: Two-dimensional anti-jamming communication based on deep reinforcement learning. In: Proc. IEEE Int. Conf. Acoust. Speech Signal Process. (ICASSP). New Orleans, LA (March 2017)
9. Kingston, D., Rasmussen, S., Humphrey, L.: Automated UAV tasks for search and surveillance. In: IEEE Conf. Control Appl. (CCA), pp. 1–8. Buenos Aires, Argentina (September 2016)
10. Li, T., Mandayam, N.B.: When users interfere with protocols: Prospect theory in wireless networks using random access and data pricing as an example. IEEE Trans. Wireless Commun. 13(4), 1888–1907 (April 2014)
11. Lu, X., Xiao, L., Dai, C., Dai, H.: UAV-aided cellular communications with deep reinforcement learning against jamming. IEEE Wireless Commun. Mag. 27(4), 48–53 (August 2020)
12. Lv, S., Xiao, L., Hu, Q., Wang, X., Hu, C., Sun, L.: Anti-jamming power control game in unmanned aerial vehicle networks. In: Proc. IEEE Global Commun. Conf. (GLOBECOM), pp. 1–6. Singapore (December 2017)
13. Marttinen, A., Wyglinski, A.M., Jantti, R.: Statistics-based jamming detection algorithm for jamming attacks against tactical MANETs. In: IEEE Military Commun. Conf., pp. 501–506. Baltimore, MD (Oct 2014)
14. Mathur, S., Reznik, A., Ye, C., Mukherjee, R., Rahman, A., et al.: Exploiting the physical layer for enhanced security. IEEE Wireless Commun. 17(5), 63–70 (Oct 2010)
15. Mnih, V., Kavukcuoglu, K., Silver, D., Rusu, A.A., Veness, J., Bellemare, M.G., Graves, A., Riedmiller, M., Fidjeland, A.K., Ostrovski, G., et al.: Human-level control through deep reinforcement learning. Nature 518(7540), 529–533 (February 2015)
16. Prelec, D.: The probability weighting function. Econometrica 66(3), 497–527 (May 1998)
17. Rappaport, T.S., et al.: Wireless communications: Principles and practice, vol. 2. Prentice Hall PTR, NJ (1996)
18. Roldán, J.J., del Cerro, J., Barrientos, A.: A proposal of methodology for multi-UAV mission modeling. In: Proc. IEEE Mediterranean Conf. Control Autom. (MED), pp. 1–7. Torremolinos, Spain (June 2015)
19. Saad, H., Mohamed, A., ElBatt, T.: Cooperative q-learning techniques for distributed online power allocation in femtocell networks. Wireless Commun. Mobile Comput. 15(15), 1929–1944 (September 2015)
20. Shahid, A., Aslam, S., Kim, H.S., Lee, K.G.: A docitive Q-learning approach towards joint resource allocation and power control in self-organised femtocell networks. Trans. Emerg. Telecommun. Technol. 26(2), 216–230 (February 2015)

21. Sheng, G., Min, M., Xiao, L., Liu, S.: Reinforcement learning-based control for unmanned aerial vehicles. J. Commun. Inf. Networks 3(3), 39–48 (October 2018)

22. Shin, H., Choi, K., Park, Y., Choi, J., Kim, Y.: Security analysis of FHSS-type drone controller. In: Proc. Int. Workshop on Inform. Security Appl., pp. 240–253. Jeju Island, Korea (August 2015)

23. Song, L., Li, Y., Han, Z.: Game-theoretic resource allocation for full-duplex communications. IEEE Wireless Commun. 23(3), 50–56 (Jun 2016)

24. Sun, R., Matolak, D.W.: Air–ground channel characterization for unmanned aircraft systems part II: Hilly and mountainous settings. IEEE Trans. Veh. Technol. 66(3), 1913–1925 (March 2017)

25. Xiao, L.: Anti-jamming transmissions in cognitive radio networks (2015). ISBN:978-3-319-24290-3

26. Xiao, L., Li, Y., Dai, C., Dai, H., Poor, H.V.: Reinforcement learning-based NOMA power allocation in the presence of smart jamming. IEEE Trans. Veh. Technol. 67(4), 3377–3389 (April 2018)

27. Xiao, L., Liu, J., Li, Q., Mandayam, N.B., Poor, H.V.: User-centric view of jamming games in cognitive radio networks. IEEE Trans. Info. Forensics Secur. 10(12), 2578–2590 (December 2015)

28. Xiao, L., Sheng, G., Liu, S., Dai, H., Peng, M., Song, J.: Deep reinforcement learning-enabled secure visible light communication against eavesdropping. IEEE Trans. Commun. 67(10), 6994–7005 (October 2019)

29. Xiao, L., Xie, C., Chen, T., Dai, H., Poor, H.V.: A mobile offloading game against smart attacks. IEEE Access 4, 2281–2291 (May 2016)

30. Xiao, L., Xie, C., Min, M., Zhuang, W.: User-centric view of unmanned aerial vehicle transmission against smart attacks. IEEE Trans. Veh. Technol. 67(4), 3420–3430 (April 2018)

31. Xiao, L., Zhuang, W., Zhou, S., Chen, C.: Learning-based VANET communication and security techniques. Springer (2019). ISBN: 978-3-030-01731-6

32. Xu, Y., et. al.: A one-leader multi-follower Bayesian-Stackelberg game for anti-jamming transmission in UAV communication networks. IEEE Access 6, 21697–21709 (April 2018)

33. Yang, Y., Park, L.T., Mandayam, N.B., Seskar, I., Glass, A.L., Sinha, N.: Prospect pricing in cognitive radio networks. IEEE Trans. Cogn. Commun. Networking 1(1), 56–70 (Mar 2015)

34. Ye, R., Xu, Q.: Learning-based power management for multicore processors via idle period manipulation. IEEE Trans. Comput. Aided Des. Integr. Circuits Syst. 33(7), 1043–1055 (Jul 2014)

35. Yu, J., Cheung, M.H., Huang, J.: Spectrum investment with uncertainty based on prospect theory. In: Proc. IEEE Int. Conf. Commun. (ICC), pp. 1620–1625. Sydney, NSW, Australia (Jun 2014)

36. Zhu, Q., Saad, W., Han, Z., Poor, H.V., Basar, T.: Eavesdropping and jamming in next-generation wireless networks: A game-theoretic approach. In: IEEE Military Commun. Conf. (MILCOM), pp. 119–124. Baltimore, MD (Nov 2011)

Visual Analysis of Adversarial Examples in Machine Learning

Wei Zong, Yang-Wai Chow, and Willy Susilo

1 Introduction

Nowadays, it is widely accepted that machine learning (ML) has demonstrated supreme capabilities in handling various practical tasks. In particular, deep neural networks (DNNs) have achieved the best results in many application domains, such as face recognition [13], automatic speech recognition (ASR) [51], and so on. ML models can be trained to learn the complicated mapping from the input domain to the output domain using massive training data. During the training process, ML models can automatically learn features of the training data, where some of these features may or may not be interpretable by humans [30]. This is a highly desirable property, as it circumvents the necessity of requiring experts to manually construct features based on their domain expertise.

However, the security of ML techniques can be compromised using adversarial attacks. The first study in this area dates back to 2004 [2], where it was shown that carefully modified spam emails could successfully fool classifiers. Years later, adversarial examples (AEs), which can fool deep learning models, were defined in the context of image recognition by Szegedy et al. [56] in 2014. The AEs were generated by adding small perturbations to the original images. From the human visual perspective, the generated AEs look indistinguishable from the original images but are misclassified by ML models. Within a short period of time following this seminal work, the study of AEs was extended to other areas of interest in ML, such as ASR [9, 67] and natural language processing (NLP) [73]. Different ML fields face different challenges when it comes to generating AEs. For example, while humans cannot perceive small perturbations in images, humans are sensitive

W. Zong · Y.-W. Chow · W. Susilo (✉)
Institute of Cybersecurity and Cryptology, School of Computing and Information Technology,
University of Wollongong, Wollongong, NSW, Australia
e-mail: wsusilo@uow.edu.au

to small perturbations in audio. Thus, the generation of audio AEs may require the incorporation of psychoacoustic techniques to make perturbations imperceptible [48].

In addition to the work on generating AEs, there have also been research efforts that have focused on defending against AEs [7, 25]. In general, defending against AEs can be seen as being more difficult than generating AEs, because failing to defend against one particular AE indicates a failure in the method of defense. One defense strategy is to reduce the effectiveness of adversarial samples by smoothening the amplitude of gradients in ML models [46]. Another strategy uses JPEG compression to destroy adversarial perturbations in the images before inputting them into ML models [12]. Other successful defense strategies against AEs rely on the detection of AEs by identifying the unique characteristics of AEs as compared to benign samples [18, 38].

Despite the powerful capabilities of ML models, it is difficult to explain the internal working of these models. Thus, a number of studies in the research community have utilized visualization as a tool for understanding ML models [27]. For example, a seminal piece of work involving ML in the image domain was done by Zeiler et al. [70], in which they projected learned features into pixel space. Their work facilitated an understanding of what was learned by each layer of a convolutional neural network (CNN). In another example, Karpathy et al. [30] visualized the activations of cells in a recurrent neural network (RNN) for NLP, where they discovered that some cells were only activated when they were processing text with specific characteristics.

Nevertheless, there is limited research that has focused on visual analysis in the context of AEs in ML. Due to the importance of visualization techniques in understanding the internal workings of ML, it would be beneficial for the research community to be aware of current progress in the visual analysis of AEs. This is the motivation behind this chapter, which presents the current state of research in this area. A summary of the research work reviewed in this chapter is presented in Table 1.

The contents of this chapter are organized as follows: Sect. 2 provides background knowledge on AEs; techniques on visualizing the generation of AEs and the properties of AEs are reviewed in Sects. 3 and 4, respectively; Sect. 5 presents work on visually distinguishing AEs; the research on visualizing the robustness of ML models against AEs is described in Sect. 6; challenges and future research directions are discussed in Sect. 7; and finally, Sect. 8 concludes the chapter.

Table 1 A summary of research efforts and their different objectives in the visual analysis of AEs

Visualization objectives	Research work
Generation of AEs	[35, 36, 45, 61, 62]
Properties of AEs	[37, 42, 53, 71]
Distinguishing AEs	[11, 15, 39, 65, 68]
Robustness against AEs	[6, 17, 32, 60, 72]

2   Adversarial Examples

This section presents background knowledge on AEs. AEs are generated by adding small perturbations to the original data in a way where classifiers are misled into producing erroneous results. Although AEs were first defined in the field of computer vision [56], interest in AEs quickly spreads to other fields, including ASR [48, 63, 67] and NLP [74].

To make AEs difficult, or even impossible, to be distinguished from original data, constraints must be imposed on the perturbations. Image AEs are usually constrained by p-norm, which is generally sufficient to make perturbations imperceptible. However, this constraint does not work for audio AEs, because humans can perceive small perturbations in audio. It is therefore deemed necessary to incorporate psychoacoustic techniques in constraining perturbations that are added to audio [48, 50]. In the NLP domain, defining appropriate constraints on perturbations is even more challenging. This is because features in NLP are discrete, so adversaries cannot apply slight imperceptible modifications as in the case of generating image or audio AEs [35]. Thus, AEs in NLP are generated through the introduction of spelling errors or replacing original words with different words. Although words can be represented by vectors in continuous space, slight modification of these vectors will produce a string of text that is not necessarily an actual word [75].

AEs can be categorized into white-box or black-box methods. White-box AEs are generated with complete knowledge of the target ML model, including its architecture, weights, and so on [4, 14, 34]. Therefore, the generation process is normally formulated as solving an optimization problem because the loss gradients are known [5, 69]. In contrast, black-box AEs are generated without knowledge of the target ML model. The only information available is input and resulting output pairs when passed to the ML model [8, 44]. The generation of black-box AEs can make use of genetic algorithms to search a large state space [1, 31, 57]. Alternatively, adversaries can train a surrogate model to generate white-box AEs. These white-box AEs can subsequently be used as black-box AEs to fool the target ML model [37, 78].

The process of generating AEs can be represented as modifying a data point until it crosses the ML decision boundaries. For example, a polyhedron can be used to approximate the decision boundaries of a classifier, and an AE is generated when the data point is moved outside of these boundaries [43]. In this manner, the generation of AEs can be defined as an iterative process. At iteration i, the polyhedron $\tilde{P}_i$ is formulated using Eq. 1 [43], where x_0 is the original image, $f : \mathbb{R}^n \to \mathbb{R}^c$ is a classifier, the kth output of $f(x)$ is $f_k(x)$, and $\hat{k} = \arg\max_x f_k(x)$. At each iteration, a greedy update is applied that calculates the perturbation vector to reach the boundary of $\tilde{P}_i$ and to update x_i accordingly.

$$\tilde{P}_i = \bigcap_{k=1}^{c} \{x : f_k(x_i) - f_{\hat{k}(x_0)}(x_i) + \nabla f_k(x_i)^\mathsf{T} x - \nabla f_{\hat{k}(x_0)}(x_i)^\mathsf{T} x \le 0\}. \tag{1}$$

3   Generation of Adversarial Examples

In this section, research studies that visualize the generation of AEs are presented. To date, there is limited research in this area, despite the fact that the visualization of how AEs fool ML models can potentially be a vital tool for researchers to understand the underlying mechanisms of AEs.

The first work in this area was conducted by Norton et al. [45]. In their work, they introduced a visualization tool called Adversarial Playground. This tool provides a web-based interface that users can use to experiment with effects of different AEs. Parameters to generate AEs are sent to a server where the intensive calculation is performed. Then, the generated AEs along with other information, such as classification likelihoods, are rendered on the client side. Adversarial Playground provides three benefits, namely, educational, interactivity, and modularity. The educational benefit is obvious, since users can gain a clear understanding of how AEs can fool ML models. The interactivity and modularity benefits refer to Adversarial Playground's quick response to users, and the fact that other researchers can easily plug Adversarial Playground into their own framework.

Other than Adversarial Playground, other works on visualizing the generation of AEs are primarily concentrated in the NLP domain. AllenNLP Interpret, proposed by Wallace et al. [62], is built on top of AllenNLP [22], which is a platform for NLP research. AllenNLP Interpret can visualize two AEs: Hotflip [16] and Input Reduction [21]. Hotflip changes the prediction of ML models by changing words from the input based on gradient information. In contrast, Input Reduction removes as many words from the input as possible, while keeping the prediction unchanged.

Alongside AllenNLP Interpret, concurrent work was conducted by Laughlin et al. [35], which also visualizes how AEs can fool NLP models. The difference is that the visualization framework in Laughlin et al. [35] is based on existing evolutionary attack strategies, rather than on Hotflip or Input Reduction. Thus, the framework is a black-box system, and it can work with any classifier as long as it outputs classification scores. In addition, the dashboard provides an interface for manually correcting the generated AEs. This functionality is useful since automatically generated AEs usually suffer from poor semantics. To help manual correction, scatterplots are provided to explore the word replacement space. Another contribution of this framework is that it can be used for testing the robustness of ML models. The robustness of different classifiers can be compared using information such as, the difference between generated AEs and the original text, the percentage of words replaced and the difference in classifier output scores.

Question answering (QA) is an important discipline within the NLP domain [64], and AEs also present security challenges in this field [29]. Wallace et al. [61] and Lee et al. [36] proposed two tools to visualize AEs in QA. Both tools support manual modification of the questions, and users can see the changed scores produced by the underlying models. The end goal is to generate questions that can fool QA models but can still be correctly answered by humans. Other than manual modification, the visualization tool proposed by Lee et al. [36] also supports automatic generation of AEs via reusable adversarial rules.

In summary, much of current research effort on visualizing the generation of AEs is mainly focused in the NLP domain. A common feature of these visualization tools is that they all support manual modification of input text for users to view changes in the resulting scores. It would be ideal for a tool to be able to automatically suggest word replacements, as done in [35]. This would significantly make the modification process easier for users. The design of these tools can also be applied to other fields, such as computer vision and ASR. For example, Adversarial Playground can be updated to support manual modification of input images and then present users with the resulting image recognition scores.

4 Properties of Adversarial Examples

In the literature, several researchers have focused on studying properties of AEs, such as universality [42] and transferability [37]. The universality property means that there exist adversarial perturbations that can be applied to any data point, while transferability of AEs refers to AEs generated using one model that can be used to fool other models. This section describes research efforts that attempt to visually explain properties of AEs.

Moosavi-Dezfooli et al. [42] present an interesting observation that there exist Universal Adversarial Perturbations (UAPs). Image AEs can be generated by directly adding these perturbations to original images. The generation of UAPs only requires a small subset of the training data. UAPs are generated by iteratively accumulating perturbations that send images to the ML decision boundaries. In their work, they provided a visual explanation as to why such universal perturbations exist [42]. Specifically, they plotted singular values of a matrix N, which is composed of normal vectors to the decision boundaries for a set of images, against a matrix with random columns. They observed that the quick decrease in singular values of N suggests that there exists a subspace that contains most of the normal vectors to decision boundaries. To verify the existence of this subspace, Moosavi-Dezfooli et al. [42] demonstrate that a random vector sampled using the first 100 singular vectors can fool nearly 38% of images. It should be noted that although UAPs were not generated directly from this subspace, the purpose of discussing and visualizing this subspace is to explain why such UAPs exist.

Another study on UAPs was recently conducted by Zhang et al. [71]. In addition to non-targeted UAPs in [42], Zhang et al. [71] proposed targeted UAPs, where perturbed images are misclassified to predetermined labels. The targeted UAPs are generated by iteratively using perturbed images to fool an ML model. To visually analyze targeted UAPs, they used a clean image, a targeted UAP, and a generated AE as separate input into an ML model. Then, pairs of corresponding elements from logit vectors were plotted. From the plots, it was clear that the Pearson correlation coefficient (PCC) between the logits of the targeted UAP and the generated AE were significantly larger than the PCC between the logits of the clean image and the generated AE. This implied that the UAP dominated the classification process, while the clean image acted like noise [71]. Based on this observation, the authors in [71] were successful in generating targeted UAPs without using the original training sets.

There has also been much research into the transferability of AEs by a number of researchers [28, 37, 59, 78]. Transferability of AEs poses a particularly severe threat to ML models since it enables black-box attacks [47]. Specifically, adversaries can train a surrogate model to generate AEs and then use these AEs to fool the target model. Seminal work visually explaining the transferability of AEs was conducted by Liu et al. [37]. In their study, five state-of-the-art models were investigated, namely, ResNet-50, ResNet-101, ResNet-152 [26], GoogLeNet [55], and VGG-16 [52]. An interesting observation from their results was that both non-targeted and targeted AEs generated using an ensemble of models, transferred well to a new model. Their results also showed that given a correctly classified image, the decision boundaries of different models were similar to each other, and as such, a generated AE could be transferred to other models.

Another property of AEs as examined by Stutz et al. [53] is whether or not AEs are on the low-dimensional data manifold. In that respect, AEs can be classified as regular or on-manifold, respectively. Regular AEs usually leave the data manifold, while on-manifold AEs are constrained to be on the data manifold. By providing a visual comparison, they showed that it was clear that perturbations of the on-manifold AE were interpretable by humans, while perturbations of the regular AE appeared to be noise-like. This result is intuitive since regular AEs are off the data manifold, while on-manifold AEs are within the data distribution. Further experiments in [53] showed that robustness of on-manifold AEs is strongly related to generalization, while robustness of regular AEs is not. Therefore, the authors argued that it is potentially possible to train models to be highly accurate and also possess an adversarial robustness property. This appears to contradict to the argument that there is a trade-off between adversarial robustness and generalization [54, 60].

To sum up, the work described in this section visually explains properties of AEs, in terms of their universality, transferability, and whether AEs are constrained on the data manifold. Visualization plays an important role in intuitively understanding these properties. Universality means there exist adversarial perturbations that can be applied to any data point, while transferability of AEs refers to AEs generated using one model that can be used to fool other models, and AEs constrained on the data manifold contain interpretable perturbations.

5   Distinguishing Adversarial Examples

This section describes research efforts that seek to visually distinguish AEs from benign samples.

Recent work conducted by Cohen et al. [11] was inspired from the fact that AEs can be visually separated from benign samples based on the correspondence of k-nearest neighbors (kNNs) and helpful samples. This correspondence can be exploited to train a simple logistic regression (LR) model for automatic detection of AEs. Specifically, a new adversarial set is generated by attacking the validation set.

For each data point in the adversarial and validation sets, the most M helpful and harmful samples from the training set are calculated by employing the influence functions proposed by Koh et al. [33], where M is a hyper-parameter. Helpful samples are those that help models to make accurate predictions, while harmful samples prevent a model from making such predictions. The nearest neighbors' ranks and distances of these samples are calculated and used to train a simple LR model. A nearest neighbors' rank of a data point is the value k that results in the data point being within the kNN range of the target. The distance of a data point is measured by the L_2 distance between the activations of this data point and the target. When a new data point is received as input, the nearest neighbors' ranks and L_2 distances of its M most helpful and harmful training data are fed into the LR to detect whether it is an AE. Their experimental results showed high accuracy in detecting six popular AEs.

Another visual method for explaining the difference between AEs and normal data is via attention heatmaps generated by class activation mapping (CAM) [77]. Attention heatmaps provide a visualization of the implicit discriminative regions of a model. This technique can intuitively explain how AEs can mislead DNNs. Recent work that compared AEs with normal examples based on attention heatmaps was conducted by Xiao et al. [65]. They proposed a novel AE, which is generated by moving pixels in the original image to new positions. Since the generation process does not modify pixel values, these AEs introduce less artifacts compared to AEs that do modify pixel values [3, 24]. As an example, they provided attention heatmaps of a normal example and its corresponding AEs. For the normal example, the attention heatmap showed that the model focused on correct regions in the image. However, for the corresponding AEs, the attention heatmaps highlighted different regions. This showed that the generated AEs misled the ML model's focus of attention. Similar results can also be found in other research, in which attention heatmaps were used to visually distinguish normal data from AEs [39, 68].

Similar to CAM, discrepancy maps (DMs) can also be used to interpret AEs [15]. A DM is generated by recording patches, which occlude parts of an image, that result in a large change in activations [76]. In other words, DMs can identify which regions of an image a neuron is looking for. Dong et al. [15] utilized DMs to interpret how AEs fool a neuron. Their experiments showed that the DMs of original images preserve semantic meanings, while DMs of AEs do not.

Overall, AEs can be visually distinguished from benign samples, even though AEs are generated by adding imperceptible perturbations. CAM and DMs are two techniques that can visually interpret which parts of images are important for classification. Benign samples usually highlight semantically meaningful regions, while AEs do not. This abnormal behavior can potentially be used to detect AEs.

6   Robustness of Models

Improving the robustness of ML models against AEs is an important topic in the literature [41, 49, 58]. This section presents a discussion on research aimed at visually explaining the robustness of models.

There have been a number of studies aimed at developing models that are robust against AEs [20, 54, 66]. Among the proposed methods, adversarial training has been shown to achieve state-of-the-art results [40]. In adversarial training, ML models are deliberately trained using AEs. However, there is a trade-off between accuracy and robustness [19]. Tsipras et al. [60] assert that this is because robust models are intrinsically different from non-robust models. They contend that non-robust models achieve high performance by relying on weakly correlated features, while features learned by robust models align well with human perception. This interesting observation has also been confirmed by other researchers [32].

Inspired by the work of Tsipras et al. [60], Chan et al. [6] proposed Jacobian adversarially regularized networks (JARNs), which improve the robustness of a neural network against AEs. The main question that it attempts to address is: due to the fact that loss gradients with respect to input images are interpretable for robust models, can we achieve robust models if we make interpretable loss gradients as a training objective? To achieve this objective, the architecture of JARN follows the same pattern as generative adversarial networks (GANs) [23]. Specifically, a generator network is trained to make the loss gradients of a classifier resemble input images by minimizing the adversarial loss. Meanwhile, a discriminator network is trained to distinguish the produced loss gradients from input images by maximizing the adversarial loss.

In their experimental results, they showed that JARN improved the robustness of the classifier. If JARN were to be trained using one-step adversarial training, the robustness could further improve. Besides visualizing the loss gradients, robustness was also explained by plotting loss surfaces of models. The visualization of loss surfaces has also been used in other studies to explain the robustness of models [17, 72].

In summary, robust ML models learn a different feature representation as compared with non-robust models [60]. This characteristic can be directly exploited to improve the robustness of models [6].

7   Challenges and Research Directions

In this section, current challenges in the field of visual analysis of AEs as well as future research directions are described.

- It is challenging to detect image AEs based on attention heatmaps. Attention heatmaps can explain which part of an image an ML model focuses on. Image AEs usually mislead a model into focusing on unimportant or even peculiar parts of an image [39]. Thus, a future direction in AEs might involve determining which parts of an image an ML model should focus its attention on and use this information to detect images that attempt to divert a model's focus of attention.

- Another challenge is in visualizing the robustness of models that handle sequence data, such as models in ASR or NLP. In the domain of image recognition, loss gradients of robust models with respect to input images are interpretable and resemble the input [60]. This may not be true for models that handle sequence data, because sequence data are fundamentally different from data in the form of static images. Although there is existing research investigating the robustness of sequence models [10], how to understand and explain the robustness of such models in a visual manner is still an open question.
- Even though much research attention has focused on analyzing image and text AEs, there is currently limited work on visually analyzing audio AEs. A potential research direction is to visualize the generation of audio AEs in a manner similar to the visualization of text AEs [62]. Research outcomes in this area may help explain which parts of audio signals are important for ML models.

8 Conclusion

The field of ML has seen rapid advancements in recent decades. Meanwhile, security threats to ML models, such as AEs, have attracted the interest of a number of researchers, as these can pose serious threats to ML models. AEs are generated by modifying original data in an imperceptible manner. This chapter has presented a review of methods that attempt to visually analyze AEs. This has included methods for visualizing the generation of AEs, the properties of AEs, how to distinguish AEs, and the robustness of ML models against AEs. Along with existing research, the current challenges and interesting future research directions in this field were also described. As such, the content in this chapter serves as a valuable reference for researchers who are interested in visually analyzing AEs.

References

1. Alzantot, M., Balaji, B., Srivastava, M.B.: Did you hear that? adversarial examples against automatic speech recognition. CoRR (2018). arXiv:abs/1801.00554
2. Biggio, B., Roli, F.: Wild patterns: Ten years after the rise of adversarial machine learning. Pattern Recognition **84**, 317–331 (2018)
3. Carlini, N., Wagner, D.: Adversarial examples are not easily detected: Bypassing ten detection methods. In: Proceedings of the 10th ACM Workshop on Artificial Intelligence and Security, pp. 3–14 (2017)
4. Carlini, N., Wagner, D.: Towards evaluating the robustness of neural networks. In: 2017 IEEE Symposium on Security and Privacy (SP), pp. 39–57. IEEE (2017)
5. Carlini, N., Wagner, D.: Audio adversarial examples: Targeted attacks on speech-to-text. In: 2018 IEEE Security and Privacy Workshops (SPW), pp. 1–7. IEEE (2018)

6. Chan, A., Tay, Y., Ong, Y., Fu, J.: Jacobian adversarially regularized networks for robustness. In: 8th International Conference on Learning Representations, ICLR 2020, Addis Ababa, Ethiopia, April 26–30, 2020. https://OpenReview.net

7. Chen, H., Zhang, H., Boning, D., Hsieh, C.-J.: Robust decision trees against adversarial examples. Preprint (2019). arXiv:1902.10660

8. Chen, P., Zhang, H., Sharma, Y., Yi, J., Hsieh, C.: ZOO: zeroth order optimization based black-box attacks to deep neural networks without training substitute models. In: Thuraisingham, B.M., Biggio, B., Freeman, D.M., Miller, B., Sinha, A. (eds.) Proceedings of the 10th ACM Workshop on Artificial Intelligence and Security, AISec@CCS 2017, Dallas, TX, USA, November 3, 2017, pp. 15–26. ACM (2017)

9. Chen, Y., Yuan, X., Zhang, J., Zhao, Y., Zhang, S., Chen, K., Wang, X.: Devil's whisper: A general approach for physical adversarial attacks against commercial black-box speech recognition devices. In: 29th USENIX Security Symposium (USENIX Security 20) (2020)

10. Cheng, M., Yi, J., Chen, P.-Y., Zhang, H., Hsieh, C.-J.: Seq2sick: Evaluating the robustness of sequence-to-sequence models with adversarial examples. In: AAAI, pp. 3601–3608 (2020)

11. Cohen, G., Sapiro, G., Giryes, R.: Detecting adversarial samples using influence functions and nearest neighbors. In: Proceedings of the IEEE/CVF Conference on Computer Vision and Pattern Recognition, pp. 14,453–14,462 (2020)

12. Das, N., Shanbhogue, M., Chen, S., Hohman, F., Chen, L., Kounavis, M.E., Chau, D.H.: Keeping the bad guys out: Protecting and vaccinating deep learning with JPEG compression. CoRR (2017). arXiv:abs/1705.02900

13. Deng, J., Guo, J., Xue, N., Zafeiriou, S.: ArcFace: Additive angular margin loss for deep face recognition. In: Proceedings of the IEEE Conference on Computer Vision and Pattern Recognition, pp. 4690–4699 (2019)

14. Dong, Y., Liao, F., Pang, T., Su, H., Zhu, J., Hu, X., Li, J.: Boosting adversarial attacks with momentum. In: 2018 IEEE Conference on Computer Vision and Pattern Recognition, CVPR 2018, Salt Lake City, UT, USA, June 18–22, pp. 9185–9193, 2018. IEEE Computer Society (2018)

15. Dong, Y., Su, H., Zhu, J., Bao, F.: Towards interpretable deep neural networks by leveraging adversarial examples. Preprint (2017). arXiv:1708.05493

16. Ebrahimi, J., Rao, A., Lowd, D., Dou, D.: Hotflip: White-box adversarial examples for text classification. Preprint (2017). arXiv:1712.06751

17. Engstrom, L., Ilyas, A., Athalye, A.: Evaluating and understanding the robustness of adversarial logit pairing. Preprint (2018). arXiv:1807.10272

18. Esmaeilpour, M., Cardinal, P., Koerich, A.L.: Detection of adversarial attacks and characterization of adversarial subspace. In: ICASSP 2020-2020 IEEE International Conference on Acoustics, Speech and Signal Processing (ICASSP), pp. 3097–3101. IEEE (2020)

19. Fawzi, A., Fawzi, O., Frossard, P.: Analysis of classifiers' robustness to adversarial perturbations. Machine Learning 107(3), 481–508 (2018)

20. Fawzi, A., Moosavi-Dezfooli, S.M., Frossard, P.: Robustness of classifiers: from adversarial to random noise. In: Advances in Neural Information Processing Systems, pp. 1632–1640 (2016)

21. Feng, S., Wallace, E., Grissom II, A., Iyyer, M., Rodriguez, P., Boyd-Graber, J.: Pathologies of neural models make interpretations difficult. Preprint (2018). arXiv:1804.07781

22. Gardner, M., Grus, J., Neumann, M., Tafjord, O., Dasigi, P., Liu, N., Peters, M., Schmitz, M., Zettlemoyer, L.: AllenNLP: A deep semantic natural language processing platform. Preprint (2018). arXiv:1803.07640

23. Goodfellow, I., Pouget-Abadie, J., Mirza, M., Xu, B., Warde-Farley, D., Ozair, S., Courville, A., Bengio, Y.: Generative adversarial nets. In: Advances in Neural Information Processing Systems, pp. 2672–2680 (2014)

24. Goodfellow, I.J., Shlens, J., Szegedy, C.: Explaining and harnessing adversarial examples. In: Bengio, Y., LeCun, Y. (eds.) 3rd International Conference on Learning Representations, ICLR 2015, San Diego, CA, USA, May 7–9, 2015, Conference Track Proceedings (2015)

25. Guo, M., Yang, Y., Xu, R., Liu, Z., Lin, D.: When NAS meets robustness: In search of robust architectures against adversarial attacks. In: Proceedings of the IEEE/CVF Conference on Computer Vision and Pattern Recognition, pp. 631–640 (2020)

26. He, K., Zhang, X., Ren, S., Sun, J.: Deep residual learning for image recognition. In: 2016 IEEE Conference on Computer Vision and Pattern Recognition, CVPR 2016, Las Vegas, NV, USA, June 27–30, pp. 770–778, 2016. IEEE Computer Society (2016)

27. Hohman, F., Kahng, M., Pienta, R., Chau, D.H.: Visual analytics in deep learning: An interrogative survey for the next frontiers. IEEE Trans. Visual. Comput. Graph. **25**(8), 2674–2693 (2018)

28. Inkawhich, N., Wen, W., Li, H.H., Chen, Y.: Feature space perturbations yield more transferable adversarial examples. In: Proceedings of the IEEE Conference on Computer Vision and Pattern Recognition, pp. 7066–7074 (2019)

29. Jia, R., Liang, P.: Adversarial examples for evaluating reading comprehension systems. In: Palmer, M., Hwa, R., Riedel, S. (eds.) Proceedings of the 2017 Conference on Empirical Methods in Natural Language Processing, EMNLP 2017, Copenhagen, Denmark, September 9–11, pp. 2021–2031, 2017. Association for Computational Linguistics (2017)

30. Karpathy, A., Johnson, J., Li, F.: Visualizing and understanding recurrent networks. CoRR (2015). arXiv:abs/1506.02078

31. Khare, S., Aralikatte, R., Mani, S.: Adversarial black-box attacks on automatic speech recognition systems using multi-objective evolutionary optimization. In: Kubin, G., Kacic, Z. (eds.) Interspeech 2019, 20th Annual Conference of the International Speech Communication Association, Graz, Austria, 15–19 September 2019, pp. 3208–3212. ISCA (2019)

32. Kim, B., Seo, J., Jeon, T.: Bridging adversarial robustness and gradient interpretability. Preprint (2019). arXiv:1903.11626

33. Koh, P.W., Liang, P.: Understanding black-box predictions via influence functions. In: Precup, D., Teh, Y.W. (eds.) Proceedings of the 34th International Conference on Machine Learning, ICML 2017, Sydney, NSW, Australia, 6-11 August 2017, Proceedings of Machine Learning Research, vol. 70, pp. 1885–1894. PMLR (2017)

34. Kurakin, A., Goodfellow, I.J., Bengio, S.: Adversarial examples in the physical world. In: 5th International Conference on Learning Representations, ICLR 2017, Toulon, France, April 24–26, 2017, Workshop Track Proceedings. https://OpenReview.net (2017)

35. Laughlin, B., Collins, C., Sankaranarayanan, K., El-Khatib, K.: A visual analytics framework for adversarial text generation. Preprint (2019). arXiv:1909.11202

36. Lee, G., Kim, S., Hwang, S.w.: QADiver: Interactive framework for diagnosing QA models. In: Proceedings of the AAAI Conference on Artificial Intelligence, vol. 33, pp. 9861–9862 (2019)

37. Liu, Y., Chen, X., Liu, C., Song, D.: Delving into transferable adversarial examples and black-box attacks. In: 5th International Conference on Learning Representations, ICLR 2017, Toulon, France, April 24–26, 2017, Conference Track Proceedings (2017)

38. Ma, X., Li, B., Wang, Y., Erfani, S.M., Wijewickrema, S., Schoenebeck, G., Song, D., Houle, M.E., Bailey, J.: Characterizing adversarial subspaces using local intrinsic dimensionality. Preprint (2018). arXiv:1801.02613

39. Ma, X., Niu, Y., Gu, L., Wang, Y., Zhao, Y., Bailey, J., Lu, F.: Understanding adversarial attacks on deep learning based medical image analysis systems. Pattern Recognition, 107332 (2020)

40. Madry, A., Makelov, A., Schmidt, L., Tsipras, D., Vladu, A.: Towards deep learning models resistant to adversarial attacks. In: 6th International Conference on Learning Representations, ICLR 2018, Vancouver, BC, Canada, April 30–May 3, 2018, Conference Track Proceedings. https://OpenReview.net (2018)

41. Mao, C., Zhong, Z., Yang, J., Vondrick, C., Ray, B.: Metric learning for adversarial robustness. In: Advances in Neural Information Processing Systems, pp. 480–491 (2019)

42. Moosavi-Dezfooli, S.M., Fawzi, A., Fawzi, O., Frossard, P.: Universal adversarial perturbations. In: Proceedings of the IEEE Conference on Computer Vision and Pattern Recognition, pp. 1765–1773 (2017)

43. Moosavi-Dezfooli, S.M., Fawzi, A., Frossard, P.: DeepFool: a simple and accurate method to fool deep neural networks. In: Proceedings of the IEEE Conference on Computer Vision and Pattern Recognition, pp. 2574–2582 (2016)

44. Nayebi, A., Ganguli, S.: Biologically inspired protection of deep networks from adversarial attacks. CoRR (2017). arXiv:abs/1703.09202
45. Norton, A.P., Qi, Y.: Adversarial-playground: A visualization suite showing how adversarial examples fool deep learning. In: 2017 IEEE Symposium on Visualization for Cyber Security (VizSec), pp. 1–4. IEEE (2017)
46. Papernot, N., McDaniel, P., Wu, X., Jha, S., Swami, A.: Distillation as a defense to adversarial perturbations against deep neural networks. In: 2016 IEEE Symposium on Security and Privacy (SP), pp. 582–597. IEEE (2016)
47. Papernot, N., McDaniel, P.D., Goodfellow, I.J.: Transferability in machine learning: from phenomena to black-box attacks using adversarial samples. CoRR (2016). arXiv:abs/1605.07277
48. Qin, Y., Carlini, N., Cottrell, G.W., Goodfellow, I.J., Raffel, C.: Imperceptible, robust, and targeted adversarial examples for automatic speech recognition. In: Proceedings of the 36th International Conference on Machine Learning, ICML 2019, 9–15 June 2019, Long Beach, California, USA, pp. 5231–5240 (2019)
49. Samangouei, P., Kabkab, M., Chellappa, R.: Defense-GAN: Protecting classifiers against adversarial attacks using generative models. In: 6th International Conference on Learning Representations, ICLR 2018, Vancouver, BC, Canada, April 30–May 3, 2018, Conference Track Proceedings. https://OpenReview.net (2018)
50. Schönherr, L., Kohls, K., Zeiler, S., Holz, T., Kolossa, D.: Adversarial attacks against automatic speech recognition systems via psychoacoustic hiding. In: 26th Annual Network and Distributed System Security Symposium, NDSS 2019, San Diego, California, USA, February 24–27, 2019. The Internet Society (2019)
51. Shen, J., Nguyen, P., Wu, Y., Chen, Z., Chen, M.X., Jia, Y., Kannan, A., Sainath, T.N., Cao, Y., Chiu, C., He, Y., Chorowski, J., Hinsu, S., Laurenzo, S., Qin, J., Firat, O., Macherey, W., Gupta, S., Bapna, A., Zhang, S., Pang, R., Weiss, R.J., Prabhavalkar, R., Liang, Q., Jacob, B., Liang, B., Lee, H., Chelba, C., Jean, S., Li, B., Johnson, M., Anil, R., Tibrewal, R., Liu, X., Eriguchi, A., Jaitly, N., Ari, N., Cherry, C., Haghani, P., Good, O., Cheng, Y., Alvarez, R., Caswell, I., Hsu, W., Yang, Z., Wang, K., Gonina, E., Tomanek, K., Vanik, B., Wu, Z., Jones, L., Schuster, M., Huang, Y., Chen, D., Irie, K., Foster, G.F., Richardson, J., et al.: Lingvo: a modular and scalable framework for sequence-to-sequence modeling. CoRR (2019). arXiv:abs/1902.08295
52. Simonyan, K., Zisserman, A.: Very deep convolutional networks for large-scale image recognition. In: Bengio, Y., LeCun, Y. (eds.) 3rd International Conference on Learning Representations, ICLR 2015, San Diego, CA, USA, May 7–9, 2015, Conference Track Proceedings (2015)
53. Stutz, D., Hein, M., Schiele, B.: Disentangling adversarial robustness and generalization. In: Proceedings of the IEEE Conference on Computer Vision and Pattern Recognition, pp. 6976–6987 (2019)
54. Su, D., Zhang, H., Chen, H., Yi, J., Chen, P., Gao, Y.: Is robustness the cost of accuracy? - A comprehensive study on the robustness of 18 deep image classification models. In: Ferrari, V., Hebert, M., Sminchisescu, C., Weiss, Y. (eds.) Computer Vision - ECCV 2018 - 15th European Conference, Munich, Germany, September 8–14, 2018, Proceedings, Part XII, Lecture Notes in Computer Science, vol. 11216, pp. 644–661. Springer (2018)
55. Szegedy, C., Liu, W., Jia, Y., Sermanet, P., Reed, S.E., Anguelov, D., Erhan, D., Vanhoucke, V., Rabinovich, A.: Going deeper with convolutions. In: IEEE Conference on Computer Vision and Pattern Recognition, CVPR 2015, Boston, MA, USA, June 7–12, 2015, pp. 1–9. IEEE Computer Society (2015)
56. Szegedy, C., Zaremba, W., Sutskever, I., Bruna, J., Erhan, D., Goodfellow, I.J., Fergus, R.: Intriguing properties of neural networks. In: Bengio, Y., LeCun, Y. (eds.) 2nd International Conference on Learning Representations, ICLR 2014, Banff, AB, Canada, April 14–16, 2014, Conference Track Proceedings (2014)
57. Taori, R., Kamsetty, A., Chu, B., Vemuri, N.: Targeted adversarial examples for black box audio systems. In: 2019 IEEE Security and Privacy Workshops (SPW), pp. 15–20. IEEE (2019)

58. Tramèr, F., Kurakin, A., Papernot, N., Goodfellow, I.J., Boneh, D., McDaniel, P.D.: Ensemble adversarial training: Attacks and defenses. In: 6th International Conference on Learning Representations, ICLR 2018, Vancouver, BC, Canada, April 30–May 3, 2018, Conference Track Proceedings. https://OpenReview.net (2018)
59. Tramèr, F., Papernot, N., Goodfellow, I., Boneh, D., McDaniel, P.: The space of transferable adversarial examples. Preprint (2017). arXiv:1704.03453
60. Tsipras, D., Santurkar, S., Engstrom, L., Turner, A., Madry, A.: Robustness may be at odds with accuracy. In: 7th International Conference on Learning Representations, ICLR 2019, New Orleans, LA, USA, May 6–9, 2019. https://OpenReview.net (2019)
61. Wallace, E., Boyd-Graber, J.: Trick me if you can: Adversarial writing of trivia challenge questions. In: ACL Student Research Workshop (2018)
62. Wallace, E., Tuyls, J., Wang, J., Subramanian, S., Gardner, M., Singh, S.: AllenNLP interpret: A framework for explaining predictions of NLP models. Preprint (2019). arXiv:1909.09251
63. Wang, Q., Guo, P., Xie, L.: Inaudible adversarial perturbations for targeted attack in speaker recognition. Preprint (2020). arXiv:2005.10637
64. Wang, W., Yang, N., Wei, F., Chang, B., Zhou, M.: Gated self-matching networks for reading comprehension and question answering. In: Proceedings of the 55th Annual Meeting of the Association for Computational Linguistics (Volume 1: Long Papers), pp. 189–198 (2017)
65. Xiao, C., Zhu, J., Li, B., He, W., Liu, M., Song, D.: Spatially transformed adversarial examples. In: 6th International Conference on Learning Representations, ICLR 2018, Vancouver, BC, Canada, April 30–May 3, 2018, Conference Track Proceedings. https://OpenReview.net (2018)
66. Xiao, K.Y., Tjeng, V., Shafiullah, N.M.M., Madry, A.: Training for faster adversarial robustness verification via inducing ReLU stability. In: 7th International Conference on Learning Representations, ICLR 2019, New Orleans, LA, USA, May 6–9, 2019. https://OpenReview.net (2019)
67. Xie, Y., Shi, C., Li, Z., Liu, J., Chen, Y., Yuan, B.: Real-time, universal, and robust adversarial attacks against speaker recognition systems. In: ICASSP 2020-2020 IEEE International Conference on Acoustics, Speech and Signal Processing (ICASSP), pp. 1738–1742. IEEE (2020)
68. Xu, K., Liu, S., Zhang, G., Sun, M., Zhao, P., Fan, Q., Gan, C., Lin, X.: Interpreting adversarial examples by activation promotion and suppression. Preprint (2019). arXiv:1904.02057
69. Yakura, H., Sakuma, J.: Robust audio adversarial example for a physical attack. In: S. Kraus (ed.) Proceedings of the Twenty-Eighth International Joint Conference on Artificial Intelligence, IJCAI 2019, Macao, China, August 10–16, 2019, pp. 5334–5341. https://ijcai.org (2019)
70. Zeiler, M.D., Fergus, R.: Visualizing and understanding convolutional networks. In: D.J. Fleet, T. Pajdla, B. Schiele, T. Tuytelaars (eds.) Computer Vision - ECCV 2014 - 13th European Conference, Zurich, Switzerland, September 6–12, 2014, Proceedings, Part I, Lecture Notes in Computer Science, vol. 8689, pp. 818–833. Springer (2014)
71. Zhang, C., Benz, P., Imtiaz, T., Kweon, I.S.: Understanding adversarial examples from the mutual influence of images and perturbations. In: Proceedings of the IEEE/CVF Conference on Computer Vision and Pattern Recognition, pp. 14,521–14,530 (2020)
72. Zhang, H., Wang, J.: Defense against adversarial attacks using feature scattering-based adversarial training. In: Advances in Neural Information Processing Systems, pp. 1831–1841 (2019)
73. Zhang, H., Zhou, H., Miao, N., Li, L.: Generating fluent adversarial examples for natural languages. In: Proceedings of the 57th Annual Meeting of the Association for Computational Linguistics, pp. 5564–5569 (2019)
74. Zhang, W.E., Sheng, Q.Z., Alhazmi, A., Li, C.: Adversarial attacks on deep-learning models in natural language processing: A survey. ACM Trans. Intell. Syst. Technol. (TIST) 11(3), 1–41 (2020)
75. Zhang, W.E., Sheng, Q.Z., Alhazmi, A.A.F.: Generating textual adversarial examples for deep learning models: A survey. CoRR (2019). arXiv:abs/1901.06796

76. Zhou, B., Khosla, A., Lapedriza, À., Oliva, A., Torralba, A.: Object detectors emerge in deep scene CNNs. In: Bengio, Y., LeCun, Y. (eds.) 3rd International Conference on Learning Representations, ICLR 2015, San Diego, CA, USA, May 7–9, 2015, Conference Track Proceedings (2015)
77. Zhou, B., Khosla, A., Lapedriza, A., Oliva, A., Torralba, A.: Learning deep features for discriminative localization. In: Proceedings of the IEEE Conference on Computer Vision and Pattern Recognition, pp. 2921–2929 (2016)
78. Zhou, W., Hou, X., Chen, Y., Tang, M., Huang, X., Gan, X., Yang, Y.: Transferable adversarial perturbations. In: Proceedings of the European Conference on Computer Vision (ECCV), pp. 452–467 (2018)

Adversarial Attacks Against Deep Learning-Based Speech Recognition Systems

Xuejing Yuan, Yuxuan Chen, Kai Chen, Shengzhi Zhang,
and XiaoFeng Wang

1 Introduction

With the rapid development of deep learning technologies, automatic speech recognition (ASR) applications have achieved excellent accuracy. Speech-to-Text API services enable developers embed flexible voice capabilities in a variety of products. Smart speakers (e.g., Amazon Echo, Google Home, etc.) and virtual voice assistant (e.g., Google Assistant and Microsoft Cortana, etc.) have become quite popular. In this paper, we use intelligent voice control (IVC) devices to refer to all the voice-enabled centralized control devices and smartphones or tablets. In our life, various smart home devices, like smart lock, smart switch, and social networks, can be connected to such voice "hub." More importantly, autopilot can be controlled by voice commands with ASR services (e.g., Amazon Auto SDK). However, researchers show that the maliciously and even slightly crafted samples (e.g., adversarial examples, AEs) can cause deep neural network (DNN) models to

X. Yuan · K. Chen (✉)
SKLOIS, Institute of Information Engineering, Chinese Academy of Sciences, Beijing, China

School of Cyber Security, University of Chinese Academy of Sciences, Beijing, China
e-mail: chenkai@iie.ac.cn

Y. Chen
Key Laboratory of Cryptologic Technology and Information Security, Ministry of Education, Shandong University, Shandong, China

School of Cyber Science and Technology, Shandong University, Shandong, China

S. Zhang
Department of Computer Science, Metropolitan College, Boston University, Boston, MA, USA

X. Wang
School of Informatics, Computing, and Engineering, Indiana University Bloomington, Bloomington, IN, USA

© The Author(s), under exclusive license to Springer Nature Singapore Pte Ltd. 2021
X. Chen et al. (eds.), *Cyber Security Meets Machine Learning*,
https://doi.org/10.1007/978-981-33-6726-5_5

misbehave in an unexpected way. In general, AE is generated by adding adversarial perturbations to a normal sample, so that it can cheat machine learning (ML) systems without human awareness. As the IVC devices are always "listening," AEs can pose serious threats to them. For example, if the attackers embed the voice commands into common audio signals (e.g., music, water sound, etc.) to generate AEs, they can attack the IVC devices by playing the AEs, while human cannot hear the commands. We propose a general approach for practical adversarial attacks against the ASR systems.

Usually, the deep learning architecture has multiple hidden layers between the input and output layers. Theoretically, sufficient number of hidden layers can model any non-linear relationship, so as to distinguish small differences between two samples. To build such a well-performing speech recognition model (e.g., Google Speech-to-Text API), a large corpus and special computing hardware are demanded in the training process. In fact, finding the vulnerability of ASR algorithms and generating AEs to attack the complex model are very difficult. In detail, there are many challenges to attack ASR systems, especially the IVC devices in the real world. (1) The perturbations should not be filtered out by IVC devices. Since most IVC devices will remove background sound to increase the accuracy of recognition, the slight perturbations should not be filtered out as background noises. (2) The AEs should work under physical world environment where there exist much more complicated impacts for the samples, e.g., electronic noise, etc. (3) The malicious command should not be noticed by ordinary users. Therefore, the perturbations should be carefully added so that the embedded command cannot attract attentions from humans. (4) Generating AEs based on a substitute model and transferability is the main method to attack black-box models. However, it is extremely difficult to simulate the large and complex black-box model, while the AEs of the common available white-box models are difficult to be transferred to complex black-box models. To achieve the practical adversarial attack, we divide our approach into three steps as follows.

"WAV-to-API" White-Box Attack We analyze the ASR algorithm of Kaldi, which is a popular open source toolkit in both academia and industry. By taking its "ASpIRE Chain Model" (referred as ASpIRE model in short) as the white-box model, we excavate the vulnerability of the algorithm and inject commands into songs, so that the generated AEs can cheat the ASpIRE model if they are directly input to it. As audios are usually saved in the form of a wave file (in "WAV" format), we take it as "WAV-to-API (WTA)" white-box attack.

"WAV-air-API" White-Box Attack We would like the attack can be succeed in the real world. Therefore, we use a "noise model" to enhance the injected command features of the AEs, so that they can practically attack the target model. We play the AE over the air with speakers and record it. If the recorded audio can be recognized as the target command by the ASpIRE model, we take the practical "WAV-air-API (WAA)" white-box attack successful.

Practical Black-Box Attack We enhance a simple local substitute model roughly approximating the target black-box model with the ASpIRE model that is more advanced yet unrelated to the target model (Google Assistant/Home, Amazon Echo, and Microsoft Cortana). So that the AEs generated based on these models have high transferability to the target black-box model. The results show that for 98% of the target commands of these devices, we can generate at least one successful AE.

2 Background and Related Work

2.1 Speech Recognition

Besides the commercial ASR systems of Google, Microsoft, Amazon, and IBM, there also exist well known open source ASR toolkits such as Kaldi, Mozilla DeepSpeech, etc. As shown in Fig. 1, the architecture of a typical ASR system includes four main procedures: pre-processing, feature extraction, acoustic model, and language model. After receiving the raw audio, the pre-processing filters out the background noises. Then, the acoustic features like Mel-Frequency Cepstral Coefficients (MFCC) are extracted and examined according to the pre-trained acoustic model to predict the most possible phonemes/characters sequence. Finally, relying on the language model, the results will be refined as a sentence with grammar rules, commonly used words, etc.

2.2 Adversarial Examples

Szegedy et al. [3] first found neural network suffers from adversarial examples (AEs). Formally speaking, one neural network can be defined as $y = F(x)$, which maps the input x to the corresponding output y. Given a specific y', the original input x, and the corresponding output y, it is feasible to find such an input x' so that $y' = F(x')$, while x and x' are too close to be distinguished by human. The above example x', together with its prediction y', is considered as targeted adversarial (TA) attack. Such attacks have potential impact since the prediction results could be manipulated by the adversary. Compared to TA attacks, untargeted adversarial (UTA) attack identifies the input x', which is still close enough to the original input x, but has different output than that of x.

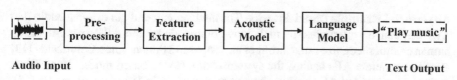

Audio Input   **Text Output**

Fig. 1 Architecture of the automatic speech recognition system

AE Attacks on Black-Box Image Recognition Models Recent researches proposed various algorithms to generate targeted AEs toward different image recognition systems. Liu et al. [4] proposed the ensemble-training approach to attack Clarifai.com, which is a black-box image classification system. Papernot et al. [5] proved that by training a local model to substitute remote DNN using the returned labels, they can attack Google and Amazon image recognition systems.

2.3 Related Work

Researchers have found that the ASR systems could be exposed to different types of attacks. We classify the existing attacks against ASR systems into four categories as below.

Speech Misinterpretation Attack The lack of proper authentication raises security and privacy concerns of ASR systems. Previous studies show third-party applications are facing misinterpretation attacks. Kumar et al. [6] presented an empirical analysis of the interpretation errors on Amazon Alexa and demonstrated that the attacker can launch a skill squatting attack. Zhang et al. [7] reported a similar attack, which utilizes a malicious skill with the similarly pronounced name to impersonate a benign skill. Zhang et al. [8] developed a linguistic-guided fuzzing tool in an attempt to systematically discover such attacks.

Signal-Manipulation Based Attacks The adversary can compromise the ASR system by either manipulating the input signal or exploiting the vulnerability of the functionalities in pre-processing. Kasmi et al. [9] found that by leveraging the intentional electromagnetic interference (IEMI) of the headset cord, voice commands can be injected into the FM signals and then can be recovered and understood by ASR systems on the smartphones. DolphinAttack [10] and SurfingAttack [11] injected voice commands into inaudible ultrasonic signal. The injected commands can be demodulated and interpreted as desired malicious commands by the target IVC devices, e.g., Apple Siri, Google Assistant, etc.

Obfuscation Based Attacks The obfuscation based attacks explore that the feature extraction of ASR systems could be manipulated. Vaidya et al. [12] and Carlini et al. [13] showed that by inverting MFCC features of the desired commands, they can get malicious audios that can be interpreted by Google Now running on a smartphone, while human beings cannot understand. Abdullah et al. [14] developed four kinds of perturbations to create malicious audio samples, so that the noisy sound can be recognized as commands by the target model.

Adversarial Example Based Attacks The attacker can craft an original audio into AEs that can be interpreted as malicious commands by the ASR systems, while human cannot recognize out. Carlini et al. proposed Hidden Voice Commands [13] attack to generate AEs against the system with a GMM-based model. In addition, their team generated AEs against the end-to-end Mozilla DeepSpeech model [15].

Schönherr et al. [16] showed that they can use psychoacoustic hiding to make imperceptible AEs toward the WSJ model of Kaldi toolkit. Qin et al. [17] succeeded in generating the imperceptible AEs to attack the Lingvo ASR system, while the results were tested in the simulated room environments. However, none of them can make the success in the real world. Furthermore, Yakura et al. [18] and Chen et al. [19] can practically attack DeepSpeech, while they cannot attack other commercial black-box ASR systems.

3 Overview

3.1 Motivation

An example for the potential security risk of adversarial attacks to the IVC system is voice navigation, which is widely used today to help drive through unfamiliar areas. Previous work [20] shows that the FM radio channel can be controlled by attackers to broadcast their malicious signals. Therefore, if the attackers craft their AE hiding a hostile navigation command and broadcast it on the selected FM radio channel, those who run voice navigation while listening to the corresponding FM channel will be impacted. This attack, if successful, will put both drivers and passengers to serious danger. Given the pervasiveness of the commercial IVC systems, it is important to understand whether such an attack is indeed realistic, particularly when the adversary has little information about how such systems work. Our research, therefore, aims at finding the answer.

White-Box Attack As DNN is widely used in the advanced ASR systems, we investigated DNN-based speech recognition algorithms and explored adversarial attacks against them. The research is motivated by the following questions: (Q1) Is it possible to build the practical adversarial attack against ASR systems, given the facts that the most ASR systems are becoming more intelligent (e.g., by integrating DNN models) and that the AEs should work in the very complicated physical environment, e.g., electronic noise from speaker, background noise, etc.? (Q2) Is it feasible to generate the AEs that are difficult, or even impossible, to be noticed by ordinary users, so the control over the ASR systems can happen in a "hidden" fashion? (Q3) If such AEs can be produced, is it possible to impact a large amount of victims in an automated way, rather than solely relying on attackers to play the AEs and affecting victims nearby?

Black-Box Attack To hack the commercial IVC devices in the real world, there are generally two requirements for the attacks: (R1) effectiveness (toward device) and (R2) concealing (toward human). Both of the two requirements emphasize the practical aspects of such attacks, that is, to deceive those devices successfully but uninterpretable by human. Unfortunately, most of the existing adversarial attacks fail either (R1) [15] or (R2) [13] in some extents. Hence, we concentrate on the research question "whether it is possible to hack those commercial IVC devices (mostly black-box based) in the real world with both (R1) and (R2) satisfied."

3.2 Technical Challenges

White-Box Attack Challenges As we totally grasp the parameters of the white-box model, we can step through the principles of speech recognition and try to find the vulnerability for generating AEs. To find *practical* AEs, a few technical challenges need to be addressed: (C1) the AE is expected to be effective in a complicated, real-world audible environment, in the presence of electronic noise from speaker and other noises; (C2) it should be stealthy, unnoticeable to ordinary users; and (C3) impactful AE should be remotely deliverable and can be played by devices from online sources, which can widely affect IVC devices.

Black-Box Attack Challenges There are two main methods to attack black-box models. Firstly, the attacker can use the "transferability" based method, i.e., AEs generated on a white-box Model A are used to attack the target Model B, as long as those two models are similar in the aspects of algorithm, training data, and model structure. Secondly, if Model A is hard to find, a local model can be trained based on the algorithm and training data to approximate Model B, to implement the "transferability." However, since Model B is black-box, the similarity is hard to determine and the algorithm as well as the training data may not be available. Therefore, such method normally suffers from the problems of uncertainty in terms of probing process and costs, especially for the commercial IVC devices whose models are quite complex for approximation.

4 White-Box Attack

We propose the "CommanderSong" attack, which aims at integrating a voice command into a clip of song. In this way, when the crafted song is played, the ASR system will recognize the injected command inside, while users are still enjoying the song as usual. In detail, based on an open source ASR platform and a white-box model, we try to discover the vulnerability of speech recognition algorithms using reverse analysis. So that we can inject voice commands into an original audio clip to generate AEs. Furthermore, we simulate the noises in the real world and enhance the robustness of the AEs, so that they can practically attack the target white-box model.

To build the "CommanderSong" attack, we leverage the Kaldi toolkit (an open source speech recognition toolkit), in which the acoustic model can be trained with DNN, and the language model is trained with Finite State Transducer (FST) algorithm. As the FST is very complex, we focus on exploiting the vulnerability of DNN. Therefore, by carefully synthesizing the outputs of the acoustic model from both the song and the given normal voice command audio, we are able to generate the AEs with minimum perturbations through gradient descent, so that the CommanderSong can be less noticeable to human users.

4.1 Threat Model of White-Box Attack

To address Q3 (Sect. 3.1), we choose songs as the "carrier" of the voice commands recognizable by ASR systems and name it as "CommanderSong" attack. The reason is at least two-fold. On one hand, enjoying songs is always a preferred way for people to relax, e.g., listening to the music station, streaming music from online libraries, or just browsing YouTube for favorite programs. Moreover, such entertainment is not restricted by using radio, CD player, or desktop computer any more. A mobile device, e.g., smartphone, allows people to enjoy songs everywhere. Hence, the attack would be widely spread. On the other hand, "hiding" the desired command in the song also makes the command much more difficult to be noticed by victims.

Actually choosing the songs as the "carrier" of desired commands makes Q2 even more challenging. Our basic idea is when generating the AEs, we revise the original song leveraging the pure voice audio of the desired command as a reference. In particular, we find the revision of the original song to generate the AEs is always a trade-off between preserving the fidelity of the original song and recognizing the desired commands from the generated AE by ASR systems. To better obfuscate the desired commands, we emphasize the former than the latter. For Q1, we use a "noise model" to simulate the real attack scenario to make sure the AEs can overcome the distortion of the noise model, so that they can be robust enough in the real world.

4.2 The Detail Decoding Process of Kaldi

There are some trained models on Kaldi website that can be used for research. As the ASpIRE model is trained with the large "Fisher" English corpus and has good recognition performance (15.6% WER), we take it as the target white-box model. After pre-processing, it extracts acoustic features (e.g., MFCC) from the raw audio. Then based on the trained probability density function (p.d.f.) of the acoustic model, those features are taken as input to DNN to compute the posterior probability matrix. The p.d.f. is indexed by the pdf identifier (pdf-id), which exactly indicates the column of the output matrix of DNN.

Phoneme is the smallest unit composing a word. There are three states (each is denoted as an HMM state) of sound production for each phoneme, and a series of transitions among those states can identify a phoneme. A transition identifier (transition-id) is used to uniquely identify the HMM-state transition. Therefore, a sequence of transition-ids can identify a phoneme, so we name such a sequence as *phoneme identifier*. Note that the transition-id is also mapped to pdf-id. Actually, during the procedure of ASpIRE model decoding, the phoneme identifiers can be obtained. By referring to the pre-obtained mapping between transition-id and pdf-id, the phoneme identifier can be mapped to a sequence of pdf-ids. Such a specific sequence of pdf-ids actually is a segment from the posterior probability matrix computed from DNN. This implies that to make the ASpIRE model decode any specific phoneme, we can have DNN to compute a posterior probability matrix containing high probabilities of the corresponding pdf-ids sequence.

To illustrate the above findings, we use the ASpIRE model to process a piece of audio with several words and obtain the intermediate results, including the posterior probability matrix computed by DNN, the transition-ids sequence, the phonemes, and the decoded words. Table 1 shows the relationship among the phoneme, HMM-state, pdf-id, transition-id, etc. analyzed from the trained acoustic model. Figure 2 demonstrates the decoded result of *Echo*, which contains three phonemes (i.e., eh_B, k_I, and ow_E). The rectangular box highlights the phoneme-id representing the corresponding phoneme, and each phoneme is identified by a sequence of transition-ids or the *phoneme identifiers*. By referring to Table 1, we can obtain the pdf-ids

Table 1 Relationship between transition-id and pdf-id

Phoneme	Phoneme-id	HMM-state	Pdf-id	Transition-id	Transition
eh_B	61	0	6383	15985	0→1
				15986	0→2
eh_B	61	1	5760	16189	Self-loop
				16190	1→2
k_I	99	0	6673	31223	0→1
				31224	0→2
k_I	99	1	3787	31379	Self-loop
				31380	1→2
ow_E	118	0	5316	39643	0→1
				39644	0→2
ow_E	118	1	8335	39897	Self-loop
				39898	1→2

Fig. 2 Result of decoding "Echo" by the ASpIRE model

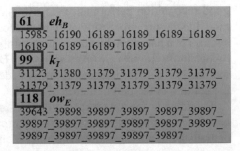

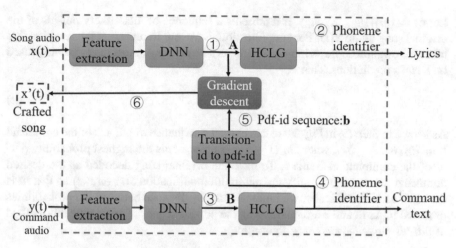

Fig. 3 Architecture of the CommanderSong WTA attack

sequence corresponding to the decoded transition-ids sequence.[1] Hence, for any posterior probability matrix demonstrating such a pdf-ids sequence can be decoded by the ASpIRE model as eh_B.

4.3 Gradient Descent to Craft Audio Clip

Based on the analysis of the ASpIRE model, we use gradient descent to generate AEs. As shown in Fig. 3, given the original song $x(t)$ and the pure voice audio of the desired command $y(t)$, we use the ASpIRE model to decode them separately. By analyzing the decoding procedures, we can get the output of DNN matrix **A** of the original song (Step ① in Fig. 3) and the phoneme identifiers of the desired command audio (Step ④ in Fig. 3).

The DNN's output A is a matrix containing the probability of each pdf-id at each frame. Suppose there are n frames and k pdf-ids, and let $a_{i,j}$ $(1 \leq i \leq n, 1 \leq j \leq k)$ be the element at the ith row and jth column in **A**. Then $a_{i,j}$ represents the probability of the jth pdf-id at frame i. For each frame, we calculate the most likely pdf-id as the one with the highest probability in that frame. That is,

$$m_i = \arg\max_j a_{i,j}. \tag{1}$$

---

[1]For instance, the pdf-ids sequence for eh_B should be *6383, 5760, 5760, 5760, 5760, 5760, 5760, 5760, 5760, 5760.*

Let $\mathbf{m} = (m_1, m_2, \ldots, m_n)$. It represents a sequence of most likely pdf-ids of the original song audio $x(t)$. For simplification, we use g to represent the function that takes the original audio as input and outputs a sequence of most likely pdf-ids based on DNN's predictions. That is,

$$g(x(t)) = \mathbf{m}. \tag{2}$$

As shown in Step ⑤ in Fig. 3, we can extract a sequence of pdf-ids of the command $\mathbf{b} = (b_1, b_2, \ldots, b_n)$, where b_i $(1 \le i \le n)$ represents the highest probability pdf-id of the command at frame i. To have the original song decoded as the desired command, we need to identify the minimum modification $\delta(t)$ on $x(t)$ so that $\mathbf{m}$ is the same as $\mathbf{b}$. Specifically, we minimize the L_1 distance between the probabilities of $\mathbf{m}$ and $\mathbf{b}$. As $\mathbf{m}$ and $\mathbf{b}$ are related with the pdf-ids sequence, we define this method as *pdf-ids sequence matching* algorithm.

$$\arg\min_{\delta(t)} \quad \|\mathbf{P}_{g(x(t)+\delta(t))} - \mathbf{P_b}\|_1. \tag{3}$$

Based on these observations, we construct the objective function as Eq. (3), where $\mathbf{P}$ is the function to represent the sum probabilities of the pdf-ids sequence. To ensure that the modified audio does not deviate too much from the original one, we optimize Eq. (3) under the constraint of $|\delta(t)| \le l$. Finally, we use gradient descent, an iterative optimization algorithm to find the local minimum of a function, to solve Eq. (3). By repeating this process until the value starts to remain stable, the algorithm is able to find a local minimum value. In particular, based on our objective function, we revise the song $x(t)$ into $x'(t) = x(t) + \delta(t)$ with the aim of making most likely pdf-ids $g\left(x'(t)\right)$ equal or close to $\mathbf{b}$. Therefore, the crafted audio $x'(t)$ can be decoded as the desired command.

4.4 Practical Adversarial Attack Against White-Box Model

In the real world, playing the AE through a speaker and recording it to physically attack the white-box model (WAV-air-API, WAA attack) are difficult. This is mainly due to the noises introduced by the speaker and environment, as well as the distortion caused by the receiver of the IVC device. Hence, our idea is to add a noise model to simulate the influence of the real world on AEs, considering the speaker noise, the receiver distortion, as well as the generic background noise, and integrate it in the approach in Sect. 4.3. We redesign the objective function as shown in Eq. (4).

$$\arg\min_{\mu(t)} \quad \|\mathbf{P}_{g(x(t)+\mu(t)+n(t))} - \mathbf{P_b}\|_1, \tag{4}$$

where $\mu(t)$ is the perturbation that we add to the original song, and $n(t)$ is the noisy audio. In this way, we can get the adversarial audio $x'(t) = x(t) + \mu(t)$ that can be used to launch the practical attack over the air.

Even though we can carefully pick up several songs and play them through the speaker and then compare the recorded audio with the original one to build a noise model, such noise model is quite device-dependent. Since different speakers and receivers may introduce different noises when playing or receiving audio, $x'(t)$ generated based on that noise model may only work with the devices that we use to capture the noise. To enhance the robustness of $x'(t)$, we introduce *random noise*, which is shown in Eq. (5). Here, the function *rand()* returns a vector of random numbers in the interval $(-N,N)$, which is saved as a "WAV" format file to represent $n(t)$. Our evaluation results show that this approach can make the adversarial audio $x'(t)$ robust enough for different speakers and receivers.

$$n(t) = rand(t), |n(t)| <= N. \tag{5}$$

4.5 Experiment Setup of CommanderSong Attack

The pure voice audio of the desired commands can be generated by any Text to-Speech (TTS) engine (e.g., Google Text-to-Speech, etc.) as long as it can be correctly recognized by the ASpIRE model. We randomly downloaded 20 songs from the Internet. To understand the impact of using different types of songs as the carrier, we intended to choose songs from different categories, i.e., popular, rock, rap, and soft music. Regarding the commands to inject, we chose 12 commonly used ones, as shown in Table 2. One GPU server (1075 MHz GPU with 12GB memory, and 512GB hard drive) was used as the computing environment.

Table 2 Results of the CommanderSong WTA attack

Command	Success rate (%)	SNR (dB)
Okay Google read mail	100	15.5
Okay Google good night	100	15.6
Okay Google flashlight on	100	14.7
Okay Google clear notification	100	14
Okay Google airplane mode on	100	16.9
Okay Google restart phone now	100	18.6
Okay Google read last sms from boss	100	15.1
Okay Google turn on wireless hot spot	100	14.7
Okay Google call one one zero one one nine one two zero	100	14.8
Echo turn off the light	100	17.3
Echo open the front door	100	17.2
Echo ask capital one to make a credit card payment	100	15.8

The WAA attack was conducted in a meeting room (16 m long, 8 m wide, and 4 m tall). We did not consider the invariance of background noise in different environments, e.g., grocery, restaurant, office, etc., due to the following reasons: (1) In a quite noisy environment like restaurant or grocery, even the original voice command $y(t)$ may not be correctly recognized by IVC devices. (2) Modeling any slightly variant background noise itself is still an open research problem. The AEs were played using three different speakers including a JBL clip2 portable speaker, an ASUS laptop, and a SENMATE broadcast equipment. The distance between the speaker and the pseudo IVC device (i.e., the microphone of the iPhone 6S) was set at 1.5 m.

4.6   Evaluation of CommanderSong Attack

SNR is a parameter widely used to quantify the level of a signal power to noise, so we use it to measure the distortion of the AE over the original audio. We use the following equation: $SNR(dB) = 10log_{10}(P_{x(t)}/P_{\delta(t)})$ to obtain SNR, where the original song $x(t)$ is the signal, while the perturbation $\delta(t)$ is the noise. Larger SNR value indicates a smaller perturbation. To evaluate the human comprehension, we conducted a survey examining the effects of "hiding" the desired command in the song. Then, we checked whether CommanderSong can spread through Internet and radio. Demos of attacks are uploaded on the website (https://sites.google.com/view/commandersong/).

WTA Attack We got more than 200 AEs and sent them to the ASpIRE model. If the AE can be identified as the command injected inside, we denote the attack is successful. Table 2 shows the WTA attack results. The success rate is calculated as the ratio of the number of words successfully decoded and the number of words in the desired command. The success rate 100% means every word in the desired command can be correctly decoded by the ASpIRE model. It can be seen that we can generate AEs for all the target commands. The high SNR (14~18.6 dB) indicates that the perturbation in the original song is very small.

WAA Attack To practically attack the white-box model, we chose two commands as in Table 2 and generated AEs based on Sect. 4.4. Then we play the AEs by speakers over the air and use iPhone 6S to record the audio, which is sent to the ASpIRE model to decode. Table 3 shows the WAA attack results. For both of the two commands, JBL speaker overwhelms the other two with the success rate up to 96%, which might indicate its sound quality is better than the other two. All the SNR values are below 2 dB, which indicates slightly bigger perturbation to the original songs due to the random noise from the signal's point of view.

Table 3 Results of the CommanderSong WAA attack

Command	Speaker	Success rate (%)	SNR (dB)
Echo ask capital one	JBL speaker	90	1.7
to make a credit	ASUS Laptop	82	1.7
card payment	SENMATE Broadcast	72	1.7
Okay Google call one	JBL speaker	96	1.3
one zero one one	ASUS Laptop	60	1.3
nine one two zero	SENMATE Broadcast	70	1.3

Table 4 Human comprehension of CommanderSong WTA AEs

Music type	Listened (%)	Abnormal (%)	Recognize command (%)
Soft music	13	15	0
Rock	33	28	0
Popular	32	26	0
Rap	41	23	0

Human Comprehension from the Survey To evaluate the effectiveness of hiding the desired command in the song, we conducted a survey[2] on Amazon Mechanical Turk (MTurk), an online marketplace for crowdsourcing intelligence. We recruited 204 individuals who were asked to listen to 26 AEs, each lasting for about 20 s (only about 4 or 5 s in the middle is crafted to contain the desired command). A series of questions regarding each audio need to be answered, e.g., (1) whether they have heard the original song before; (2) whether they heard anything abnormal than a regular song (The four options are *no*, *not sure*, *noisy*, and *words different than lyrics*.); (3) if choosing *noisy* option in (2), where they believe the noise comes from, while if choosing *words different than lyrics* option in (2), they are asked to write down those words, and how many times they listened to the song before they can recognize the words.

Table 4 shows the results of the human comprehension of our WTA AEs. We show the average results for songs belonging to the same category. Generally, the songs in the soft music category are the best candidates for the carrier of the desired command, with as low as 15% of participants noticed the abnormality. None of the participants could recognize any word of the desired command injected in the AEs of any category. Table 5 demonstrates the results of the human comprehension of our WAA AEs. On average, 40% of the participants believed the noise was generated by the speaker or like radio, while only 2.2% of them thought the noise from the AEs. In addition, less than 1% believed that there were other words except the original

---

[2]The surveys in our CommanderSong attack and Devil's Whisper attack will not cause any potential risks to the participants (physical, psychological, social, legal, etc.). The questions in our survey do not involve any confidential information about the participants. We obtained the IRB Exempt certificates from our institutes.

Table 5 Human comprehension of the CommanderSong WAA AEs

Song name	Listened (%)	Abnormal (%)	Noise speaker (%)	Noise song (%)
Did You Need It	15	67	42	1
Outlaw of Love	11	63	36	2
The Saltwater Room	27	67	39	3
Sleepwalker	13	67	41	0
Underneath	13	68	45	3
Feeling Good	38	59	36	4
Average	19.5	65.2	40	2.2

lyrics. However, none of them successfully identified any word even repeating the songs several times.

Automated Spreading To explore the potential channels that can be leveraged to impact a large amount of victims automatically, we consider the online sharing platforms like YouTube and radio signal to spread CommanderSong attack. We picked up one five-second WAA AE embedded with the command "*open the door*" and applied Windows Movie Maker software to make a video and upload it on YouTube. We then connected our desktop audio output to Bose Companion 2 speaker and installed *iFLYTEK Input* application on LG V20 smartphone. The distance between the speaker and the phone can be up to 0.5 m, and *iFLYTEK Input* can decode the command successfully. In addition, we used HackRF One, a hardware that supports Software Defined Radio (SDR) to broadcast the AE at the frequency of FM 103.4 MHz, simulating a radio station. We setup a radio at the corresponding frequency, so it can receive and broadcast the AE signal. The results show that *iFLYTEK Input* can successfully recognize the command "*open the door*" from the audio played by the radio.

5 Black-Box Attack

After successfully attacking the white-box model in the real world, we consider to attack the black-box ASR models, especially the commercial IVC devices. As we have no knowledge of the internals of the black-box ASR systems, e.g., model parameters or hyperparameters, we try the "Transferability Based Approach" (i.e., enhanced "CommanderSong" attack) and "Local Model Approximation Approach" (i.e., training a substitute model) to attack them. However, neither of them work well for all the target commands. Finally, by taking advantage of these approaches, we propose the "Alternate Models based Generation Approach" (i.e., "Devil's Whisper" attack). The results show that "Devil's Whisper" is a general approach for physical adversarial attacks against the black-box ASR systems (e.g., Google Assistant/Home, Microsoft Cortana, and Amazon Echo).

5.1   Threat Model of Black-Box Attack

A straightforward method to attack IVC devices is to generate AEs based on a white-box model and transfer the AEs to the black-box model. The success of the transferability based attack depends on the similarity between the internal structure and parameters of the white-box and target black-box models. We tested the AEs of the CommanderSong WAA attack on the ASR systems such as the Google Cloud Speech-to-text API and found that only few AEs can be partially recognized as "Hey Google," "phone," etc. The success rate of the transferability on Amazon Transcribe, the API service offered by Amazon, is even lower. Therefore, we try to enhance the transferability of CommanderSong attack and probe the target black-box model to generate robust AEs.

To probe the black-box model and train a local substitute model, we should continually input samples and get the recognized results. Even though previous work shows that we can obtain the recognized results of the IVC devices (e.g., Amazon Echo), we cannot get the confidence value, which indicates the reliability of the results. Therefore, it is very difficult to directly probe and train a substitute model of the IVC devices. Instead, we assume the corresponding online Speech-to-Text API services, i.e., providing real time decoding results and confidence values of the audio, are open to public. This assumption is valid for most of the popular IVC devices available on the market, e.g., Google Cloud Speech-to-Text API, Microsoft Bing Speech Service API, etc. Either free or pay as you go, they are accessible to the third-party developers. We assume that for the same company, the ASR system used to provide speech API service and that used for the IVC devices is the same or similar.

Once the AE is generated based on the common white-box model (e.g., ASpIRE model) and the special substitute model, we assume it will be played by speakers, which is placed not quite far away (e.g., 5 $\sim$ 200 cm) from the target IVC devices. Furthermore, we do not have the knowledge of the speakers or the microphones of the target devices. Once the attack is successful, an indicator could be observed. For instance, the AE with the command of "Echo turn off the light" is successful by observing the corresponding light off.

5.2   Transferability Based Approach

Recent research demonstrates that the transferability could work on heterogeneous models through the improvement of AE generation algorithm [5]. The momentum method [21] can accumulate a velocity vector in the gradient direction during iterations. In each iteration, the gradient is saved and then combined using a decay factor with the previously saved gradients. By combining these gradients together, the gradient direction will be much more stabilized and the transferability of AEs could be enhanced. Therefore, we enhance the CommanderSong attack by applying the Momentum based Iterative Fast Gradient Method (MI-FGM), which we name as the enhanced CommanderSong attack. Furthermore, we added random noise model in each iteration to improve the robustness of the AEs.

Based on our evaluation, the enhanced approach helps to generate a few AEs attacking black-box ASR API services (e.g., Google Cloud Speech-to-Text API) with low success rate and works even poorer on IVC devices. The main reason is that these AEs mainly depend on the example's transferability to other black-box systems. Thus, we consider the transferability based approach has one major limitation: the crafted AEs are generated more toward the white-box model. However, the decision boundaries may vary between the white-box model used to generate the AEs and the target black-box model.

5.3   Local Model Approximation Approach

Training Set with Limited Number of Phrases Generally, the commercial ASR systems are trained with significantly large proprietary corpus, and the structure of the DNN can be quite complicated. Therefore, it is extremely difficult to obtain the corpus or even infer the details about the DNN. In other words, training a local substitute model completely approximating the target commercial black-box system is almost unpractical. However, since our ultimate goal is to hack the IVC devices and in turn leverage it to compromise the victim's life, we are only interested in a limited number of commands such as "open my door," "play music," etc. A side product of selecting those commands is that, based on our experiences, the IVC devices are trained to be quite robust to them, e.g., "open my door" on Amazon Echo and "play music" on Google Assistant. Hence, we just need to train a local model approximating the target system on the most frequently used commands, also the ones we are interested in, on IVC devices.

Training Set Augment We use TTS services to generate audio clips for our desired commands as the training data for local model approximation. For the local partial approximation, the training data has two problems: the number of commands in the training data is too limited for training and the robustness of an IVC device to a command is unknown. To solve these problems, we augment the training data by tuning the TTS audio clips, i.e., either changing the speech rate or adding noises to them. We assume that the speech recognition mechanisms of the IVC devices are similar to that of the API service provided by the same company.[3] Hence, we query the corresponding online speech API service on the tuned audios and filter out those either not correctly decoded or decoded correctly but with low confidence values. The magnitude of the corpus augmented in this way is not very big, usually $3 \sim 6$

---

[3]Although previous studies [6, 7] show that it is possible to recover Speech-to-Text functionality from some IVC devices like Echo, their approaches cannot obtain the confidence values for the decoded results, which are required in our approach.

hours for ten selected commands, which can be finished in about 1500 queries on the target speech API service.

5.4 Alternate Models Based Generation Approach

Based on the substitute model of Google Speech-to-Text "command_and_search model," we use the "pdf-ids sequence matching algorithm" and "MI-FGM" to generate AEs for the target commands. Less than 25% of the commands can be successfully injected into songs to attack Google. In other words, the "Local Model Approximation Approach" has poor performance on the desired commands. As the AEs generated from the ASpIRE model can be transferred to the target black-box model in some extent, we take it as the large base model and use it to enhance the small substitute model to generate AEs. In the hope that the large base model can generate most of the acoustic features of the desired command, and the small substitute model fine-tunes them, since it was trained based on an augmented corpus (details in Sect. 5.3) that can be recognized as the target commands by the black-box model.

As shown in Fig. 4, our Devil's Whisper attack starts by transferability based approach (Step ① in Fig. 4), via an enhancement of our CommanderSong attack (details in Sect. 5.2). Then we describe the novel approach of "Alternate Models based Generation Approach" (Step ②, ③, and ④ in Fig. 4). Furthermore, the last generated AE of the base model will be fed into the substitute model (Step ② in Fig. 4). Thus, the unique features of the desired command on the target model can be adjusted in a fine-grained manner by the substitute model (Step ③ in Fig. 4). During the AEs generation process under each model, we intermittently select the crafted audios to query the target ASR API service to reduce the query times (Step ⑤ and Step ⑥ in Fig. 4). If none of them works, the last crafted audio from the substitute model will be fed to the base model as the input for the next epoch (Step ④ in Fig. 4). Finally, we select the effective AEs to compromise the IVC devices (Step ⑦ in Fig. 4).

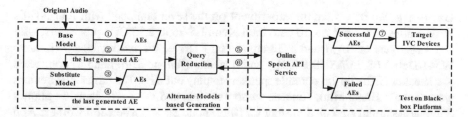

Fig. 4 Architecture of the general adversarial attack (Devil's Whisper attack) against ASR API services and IVC devices

---

Algorithm 1: Alternate Models based Generation Algorithm (Devil's Whisper)

Require: The function to get output from black-box model f_{target}, the original audio x_0, the
   generated audio x^*, the target label y ($f_{target}(x_0) \neq y$), the maximum allowed epoch
   $EpochMax$, the total iterations of the base model in one Epoch T_B, the total iterations of
   the substitute model in one Epoch T_S.

Ensure: A set of successful AEs collection X^*, all with label y under f_{target}.

 1: $x^* = x_0$; $CurrentEpoch = 0$;
 2: **while** $CurrentEpoch < EpochMax$
 3: AEgeneration (base model); if $f_{target}(x^*) = y$, put x^* into X^*;
 4: AEgeneration (substitute model); if $f_{target}(x^*) = y$, put x^* into X^*;
 5: $CurrentEpoch + +$;
 6: **end while**
 7: **return** X^*

The alternate models based generation approach (Devil's Whisper attack) is summarized in Algorithm 1. Specifically, Line 3 and Line 4 are for the AE generation on the large base model and the small substitute model, respectively. The $EpochMax$ parameter restrains the number of total alternations. During the first epoch, if after T_B iterations, effective AE is still not generated (i.e., no x^* can be recognized as the target command y), we assume the transferability based attack does not work well toward the target black-box system. We will use the output x^* from the last iteration as the input to the local substitute model and continue to generate x^* using the same gradient algorithm. If no success AEs are generated until T_S iterations with the substitute model, we input the last x^* to the base model for the next epoch. As the successful sample to attack the target ASR API service does not necessarily indicate the success toward IVC devices, instead of breaking the function, once a successful AE (x^*) is found, we save it toward the target ASR API service. Finally, we can return a set X^*, where we preserve all potential effective AEs toward the target IVC devices.

5.5 Experiment Setup of Devil's Whisper Attack

Hardware We conducted the experiment on the sever equipped with four Nvidia Tesla K40m GPUs and 2 x 10 core Intel Xeon E5-2650 2.30 GHz processors, with 131 Gigabytes of RAM and 1 Terabyte Hard Drive. We used a laptop (Lenovo W541/Dell XPS 15/ASUS P453U) and a phone (iPhone SE/iPhone 8) connected to a speaker (JBL clip 2/3 portable speaker) to play out AEs. The target IVC devices were Google Home Mini, Amazon Echo 1st Gen, and voice assistants on phones (Google Assistant App on Samsung C7100/iPhone SE and Microsoft Cortana App on Samsung C7100/iPhone 8). The AEs on IBM WAA tests were recorded by Huawei P30.

Target Models and Commands For the target black-box platforms, we considered the speech recognition devices/systems from Google, Microsoft, Amazon, and IBM.

We only focused on the specific commands frequently used on these devices, e.g., "turn off the light," "navigate to my home," "call my wife," "open YouTube," "turn on the WeMo Insight," etc. For each commercial company, we selected 10 such commands (Table 9 in Appendix) and further appended the default wake-up words for different systems (Google Assistant/Home, Amazon Echo, and Microsoft Cortana) before each of them.

Training the Local Model We chose the Mini Librispeech model[4] as the substitute model to approximate the target black-box model. We used 5 TTS services to synthesize the desired commands audio clips, with 14 speakers in total. After that, we enriched the TTS corpus by adding background noise and twisting the audio. It may lead to overfitting if the substitute model is only trained with the tuned TTS audios. Therefore, we constructed the training corpus by combining the tuned TTS audio clips (generated from the queries on the target model) and the supplemental corpus of Mini Librispeech.

The Original Audio We selected the 10 songs in the soft and popular categories of CommanderSong, which are less noisy, allowing the integrated perturbations more likely to overwhelm the background music and be decoded correctly by the target IVC devices. To further evaluate the 10 songs, we utilized two commands "Okay Google navigate to my home" and "Hey Cortana turn off the bedroom light" and ran our Devil's Whisper approach to embed the commands into the songs, against the speech recognition APIs provided by Google and Microsoft Bing. The results show that all the 10 songs can serve as carriers for the commands to ensure their recognition by the APIs. However, when listening to these AEs, we found that four instances using soft songs and one using a popular song were less stealthy than the other 5 manipulated songs and therefore selected the latter for all follow-up experiments. Our experimental results show that for each target command, there are at least 2 music clips across these 5 songs that can be crafted as effective and stealthy AEs. Therefore, we chose the 5 songs as our carriers.

5.6   Evaluation of Devil's Whisper Attack

We evaluate the effectiveness of Devil's Whisper attack on the commercial Speech-to-Text API services and IVC devices. Specifically, the effectiveness of our approach for the target commands is evaluated using the success rate of command (SRoC), that is, *the number of successful commands* vs. *the total number of the commands* evaluated on a target service. Here a successful command is the one for which we can generate at least one workable AE using our approach. Furthermore, to evaluate the robustness of our attack, we define the success rate of AE (SRoA) as the ratio of *the number of successful tests* to *the total number of tests* if an AE

---

[4]Both Mini Librispeech and ASpIRE use chain model. We used the default architecture and hyperparameters of Mini Librispeech to train all four substitute models.

Table 6 The overall SRoC results on API services of Devil's Whisper attack

	Google		Microsoft	Amazon	IBM
Black box	Phone	Command	Bing	Transcribe	STT
SRoC	10/10	10/10	10/10	4/10	10/10
SNR (dB)	11.97	9.39	13.36	11.21	10.06

Note: (1) "Phone" and "Command" represent the "phone_call model" and "command_and_search model" of Google Cloud Speech-to-Text API, respectively. (2) "Microsoft Bing" represents the Microsoft Bing Speech Service API. (3) "IBM STT" represents the IBM Speech-to-Text API. (4) The results were all based on the tests conducted in October 2019

has been repeatedly played. Attack demos are available on the website (https://sites. google.com/view/devil-whisper), and the source code is uploaded (https://github. com/RiskySignal/Devil-Whisper-Attack).

Speech-to-Text API Services Attack We feed our AEs directly into the corresponding API services and observe the results. Since we do not need to wake up the IVC devices, we consider the AE successfully attacks the target if the returned transcript matches the desired command. The results are shown in Table 6,[5] with the SNR being the average of all commands on each black-box platform (Table 8 in Appendix shows the result of each individual command). It can be seen that the attack achieves a SRoC of 100% for all Speech-to-Text API services except Amazon Transcribe API.

As for Amazon Transcribe API service, we only crafted successful AEs on 4 out of 10 target commands. We then performed more tests on Amazon Transcribe API and found that it cannot even recognize some plain TTS audio clips for the target commands correctly. In contrast, these commands can always be recognized by Amazon Echo. There can be reasons for such difference. First, different models could be used by Amazon Transcribe API and Amazon Echo. Second, the developers of Amazon Echo may set lower threshold to identify voice commands, and thus, it is more sensitive to the voice commands when used physically.

IVC Devices Attack We selected the AEs that can successfully attack the API services with high confidence score (≥ 0.6) to attack the IVC devices. Specifically, since the AEs work poorly on Amazon Transcribe API, we tested them on Amazon Echo directly, even if they failed on Amazon Transcribe API. If the devices respond to the played AE in the same way as the regular voice command from human, we consider the AE for this command successful. As we cannot find any IVC devices for IBM, we used "Wav-air-API" (WAA) to simulate it.

As shown in Table 7, the average SRoC over all IVC devices can achieve to 98%, which shows Devil's Whisper is very effective in attacking real-world IVC devices. For most of the black-box models, we can always find the AEs that successfully

---

[5]For the four models of Google Cloud Speech-to-Text API, we show the results of "phone_call model" and "command_and_search model," since according to our tests the former is similar to "video model" and the latter is similar to "default model".

Table 7 The overall SRoC results on IVC devices of Devil's Whisper attack

Black box	Google Assistant	Home	Microsoft Cortana	Amazon Echo	IBM WAA
SRoC	10/10	9/10	10/10	10/10	10/10
SNR (dB)	9.03	8.81	10.55	12.10	7.86

Note: (1) "WAA" is used to represent "Wav-air-API" attack. (2) The results were all based on the tests conducted in October 2019

attack their corresponding IVC devices from the ones that have fooled the ASR API services, except Amazon Transcribe API. The full list of target commands on different IVC devices is shown in Table 9 in Appendix. As we can see, some of them can cause safety or privacy issues, e.g., "Echo open my door," "Okay Google navigate to my home," "Okay Google take a picture," etc. The SRoA is measured over 30 tests for each target command. The results show that 76% (38/50) of the commands have SRoAs over 1/3, showing that our attack is quite robust against the IVC devices in the real world.

We used a "SMART SENSOR AS824" to measure the volume of the played AEs. The background noise was about 50 dB, and the played audios were about 65 ~ 75 dB, compared to some special cases of the sound level, e.g., talking at 3 feet (65 dB), living room music (76 dB). We also conducted experiments to test our AEs in realistic distance. For example, the AE with the command "Echo turn off the light" can successfully attack Echo as far as 200 cm away, and the AE with the command "Hey Cortana open the website" can successfully attack Microsoft Cortana as far as 50 cm away.

Human Perception of Devil's Whisper Attack We chose eight samples carrying different target commands (two AEs for Google Assistant, Google Home, Amazon Echo, and Microsoft Cortana, respectively) from our attack to conduct a survey on Amazon Mechanical Turk for the evaluation of human perception. The survey results show that 16.1% of participants think that somebody is talking in the background when they listen to Devil's Whisper, but nobody could recognize any command when an AE was played to them. Even if the participants were exposed to the same AEs for the second time, only 1.4% of them could tell over 50% of words in the target commands in the AEs. This indicates that our AEs remain very stealthy for the survey users.

6 Defense

To defense the adversarial attack against the ASR systems, we can mitigate the adversarial perturbations by signal processing (e.g., audio downsampling, signal smoothing, etc.) and authentication (e.g., source identification).

Audio Downsampling To reduce the voice command features in the AE, we can always first downsample it to a lower sampling rate and upsample it to the sampling rate that is accepted by the target black-box model. During such process, the adversarial perturbations may be mitigated, which makes the AEs fail to be recognized by the target model. For instance, we choose the AEs of IBM WAA attack. When first downsampled to 5200 Hz and then upsampled to 8000 Hz, none of them can succeed to attack IBM. In contrast, the regular recorded human voice and TTS audios can still be recognized correctly after such process. Hence, audio downsampling could be one way in detecting speech AEs.

Signal Smoothing Since the effectiveness of our AEs is highly dependent on the added perturbations by gradient algorithm, we can conduct local signal smoothing toward AEs to weaken the perturbations. Specifically, for a piece of audio x, we can replace the sample x_i with the more smooth value according to its local reference sequence, i.e., the average value of the k samples before and after x_i. Hence, the added perturbations may be mitigated by this method.

Audio Source Identification Audio source identification aims to identify the source of the audio, e.g., from an electronic speaker or human. Such defence is based on the assumption that the legitimate voice commands should only come from human rather than an electronic speaker. Therefore, if the audio is detected not from human, the audio signal will be simply ignored. Chen et al. [22] shows that they can identify the audio source by either examining the electromagnetic wave from the audio or training a model to label the audio. Such defence mechanism could work for most of the existing speech AEs that require a speaker to play. However, the attacker could play the samples over a long range, which might evade the detection.

7 Conclusion

In this paper, we perform practical adversarial attacks on ASR systems. By injecting "voice" commands into songs, our generated AEs can successfully fool the acoustic model and language model utilized in ASR systems after bypassing their pre-processing and feature extraction procedures. In addition, the AEs are stealthy enough to be recognized by humans. In detail, with no prior knowledge of the targets' machine learning models and their parameters, we enhance a simple substitute model roughly approximating the target black-box platform with a white-box model that is more advanced yet unrelated to the target. The two models are found to effectively complement each other for generating highly transferable and generic AEs on the target, which only require around 1500 queries on remote services to ensure a nearly 100% success rate of command on attacking most popular commercial ASR systems (e.g., Google Speech-to-Text, IBM Speech-to-Text, and Microsoft Bing Speech service) and IVC devices (Google Assistant, Google Home, Amazon Echo, and Microsoft Cortana).

Appendix

The detail results of *Devil's Whisper* attack on Speech-to-Text API and IVC devices are shown in Tables 8 and 9, respectively. All tests were conducted in October 2019. We mark the failure of the target command with "*X*," and the SNR of one success AE of the target command is listed. The results in Table 8 show that attacking Amazon Transcribe API is difficult. As we filter the TTS audios for the corpus of the substitute model, we find that Amazon Transcribe API is harder to recognize the TTS than other API services, such as the word "Echo." Therefore, we think that the recognition of Amazon Transcribe API is much rigorous. The practical IVC devices tests were conducted in two meeting rooms about 12 and 20 square meters, 4 m high. The volume of AEs is about 65 ~ 75 dB. The distance ranges 5 ~ 50 cm (5 ~ 200 cm for Echo). In the tests, the language of the devices needs to be English (US) only and the region/location needs to be US only (if apply). The AE carrying "Ok Google turn on the Wi-Fi" was tested on iPhone 8 using Dell XPS 15 connected to a JBL Clip 3 speaker, while it cannot succeed on iPhone SE as the other AEs.

Table 8 Detail results of the Speech-to-Text API services attack

Black-box	Command	SNR (dB)
Google phone_call API	Okay Google turn off the light	14.32
	Okay Google play music	15.17
	Okay Google take a picture	13.92
	Okay Google call 911	12.82
	Okay Google turn on airplane mode	11.91
	Okay Google navigate to my home	14.28
	Okay Google set an alarm on 8 am	12.40
	Okay Google open YouTube	7.19
	Okay Google turn on the Wi-Fi	8.21
	Okay Google turn on the Bluetooth	9.44
Google command_ and_search API	Okay Google turn off the light	13.13
	Okay Google play music	10.07
	Okay Google take a picture	9.11
	Okay Google call 911	12.80
	Okay Google turn on airplane mode	8.01
	Okay Google navigate to my home	13.36
	Okay Google set an alarm on 8 am	5.82
	Okay Google open YouTube	8.46
	Okay Google turn on the Wi-Fi	5.99
	Okay Google turn on the Bluetooth	7.11

(continued)

Table 8 (continued)

Black-box	Command	SNR (dB)
Microsoft Bing Speech Service API	Hey Cortana send a text	14.30
	Hey Cortana make it warmer	14.97
	Hey Cortana open the website	13.4
	Hey Cortana where is my phone?	13.52
	Hey Cortana what is the weather?	14.45
	Hey Cortana turn off the computer	14.11
	Hey Cortana turn on the coffee maker	13.72
	Hey Cortana turn off the bedroom lights	13.55
	Hey Cortana set the temperature to 72 degrees	9.73
	Hey Cortana add an appointment to my calendar	11.85
Amazon Transcribe API	Echo play music	12.25
	Echo call my wife	X
	Echo open my door	X
	Echo where is my car?	10.92
	Echo turn off the light	X
	Echo clear notification	13.27
	Echo what is the weather?	X
	Echo turn off the computer	8.39
	Echo turn on the TV	X
	Echo turn on the WeMo Insight	X
IBM Speech-to-Text API	Education is provided by schools	12.51
	Teachers are trained in normal schools	13.72
	What would you recommend?	12.30
	The economist provides news and information	11.54
	Business is the activity of making money	13.86
	Share the new version	11.28
	This article is about the profession	8.08
	All governments have an official form	6.10
	Children are divided by age groups into grades	6.75
	A partnership is a business owned by two or more people	4.41

Table 9 Detail results of the IVC devices attack

Black-box	Command	SNR (dB)	SRoA	Speaker	Recorder	Audio source
Google Assistant Version-0.1.18794 5513	Okay Google call 911	9.50	19/30	JBL Clip 2	iPhone SE	Lenovo W541
	Okay Google set an alarm on 8 am	8.08	4/30			
	Okay Google take a picture	5.85	5/30			
	Okay Google turn off the light	10.75	16/30			
	Okay Google play music	11.62	8/30			
	Okay Google turn on airplane mode	8.30	19/30			
	Okay Google navigate to my home	12.02	18/30			
	Okay Google open YouTube	9.49	4/30			
	Okay Google turn on the Bluetooth	9.44	15/30			
	Okay Google turn on the Wi-Fi	5.27	12/30	**	**	**
Google Home Version-1.42.171 861	Okay Google play music	11.62	28/30	JBL Clip 3	Google Home Mini	iPhone SE
	Okay Google turn off the light	10.75	15/30			
	Okay Google turn on airplane mode	8.30	18/30			
	Okay Google call 911	12.79	25/30			
	Okay Google set an alarm on 8 am	X	X			
	Okay Google take a picture	5.85	24/30			
	Okay Google navigate to my home	7.62	26/30			
	Okay Google open YouTube	9.49	22/30			
	Okay Google turn on the Wi-Fi	5.16	6/30			
	Okay Google turn on the Bluetooth	7.67	21/30			

(continued)

Table 9 (continued)

Black-box	Command	SNR (dB)	SRoA	Speaker	Recorder	Audio source
Microsoft	Hey Cortana send a text	11.71	21/30	JBL	iPhone	ASUS
Cortana	Hey Cortana make it warmer	9.28	18/30	Clip 2	8	P453U
Version-3	Hey Cortana open the website	12.44	29/30			
.3.2.2682	Hey Cortana where is my phone?	11.67	6/30			
	Hey Cortana what is the weather?	9.92	15/30			
	Hey Cortana turn off the computer	10.07	7/30			
	Hey Cortana turn on the coffee maker	10.73	15/30			
	Hey Cortana turn off the bedroom lights	9.63	13/30			
	Hey Cortana set the temperature to 72 degrees	10.24	9/30			
	Hey Cortana add an appointment to my calendar	9.77	14/30			
Amazon	Echo play music	13.43	21/30	JBL	Echo	ASUS
Echo	Echo call my wife	10.86	17/30	Clip 2	1st gen	P453U
Version-6	Echo open my door	11.36	17/30			
47588720	Echo where is my car?	11.31	23/30			
	Echo turn off the light	12.36	28/30			
	Echo clear notification	12.45	10/30			
	Echo what is the weather?	11.13	30/30			
	Echo turn off the computer	14.28	11/30			
	Echo turn on the TV	11.56	6/30			
	Echo turn on the WeMo Insight	12.21	14/30			

IBM (WAA)		4/30	JBL Clip 2	Huawei P30	iPhone SE
Education is provided by schools	9.21	4/30			
Teachers are trained in normal schools	13.74	10/30			
What would you recommend?	12.24	25/30			
The economist provides news and information	8.07	24/30			
Business is the activity of making money	4.07	24/30			
Share the new version	7.89	12/30			
This article is about the profession	7.82	26/30			
All governments have an official form	5.33	13/30			
Children are divided by age groups into grades	6.55	18/30			
A partnership is a business owned by two or more people	3.72	2/30			

References

1. Yuan, X., Chen, Y., Zhao, Y., Long, Y., Liu, X., Chen, K., xZhang, K., Huang, H., Wang, X., Gunter, C.A.: Commandersong: A systematic approach for practical adversarial voice recognition. In: 27th USENIX Security Symposium (USENIX Security 18), pp. 49–64 (2018)
2. Chen, Y., Yuan, X., Zhang, J., Zhao, Y., Zhang, S., Chen, K., Wang, X.: Devil's Whisper: A general approach for physical adversarial attacks against commercial black-box speech recognition devices. In: 29th USENIX Security Symposium (USENIX Security 20) (2020)
3. Szegedy, C., Zaremba, W., Sutskever, I., Bruna, J., Erhan, D., Goodfellow, I., Fergus, R.: Intriguing properties of neural networks. Preprint (2013). arXiv:1312.6199
4. Liu, Y., Chen, X., Liu, C., Song, D.: Delving into transferable adversarial examples and black-box attacks. Preprint (2016). arXiv:1611.02770
5. Papernot, N., McDaniel, P., Goodfellow, I., Jha, S., Celik, Z.B., Swami, A.: Practical black-box attacks against machine learning. In: Proceedings of the 2017 ACM on Asia Conference on Computer and Communications Security, pp. 506–519. ACM (2017)
6. D. Kumar, R. Paccagnella, P. Murley, E. Hennenfent, J. Mason, A. Bates, M. Bailey. Skill squatting attacks on Amazon Alexa. In: 27th {USENIX} Security Symposium ({USENIX} Security 18), pp. 33–47 (2018)
7. Zhang, N., Mi, X., Feng, X., Wang, X., Tian, Y., Qian, F.: Dangerous skills: Understanding and mitigating security risks of voice-controlled third-party functions on virtual personal assistant systems. In: 2019 IEEE Symposium on Security and Privacy (SP), pp. 1381–1396. IEEE (2019)
8. Zhang, Y., Xu, L., Mendoza, A., Yang, G., Chinprutthiwong, P., Gu, G.: Life after speech recognition: Fuzzing semantic misinterpretation for voice assistant applications. In: Network and Distributed Systems Security (NDSS) Symposium (2019)
9. Kasmi, C., Esteves, J.L.: IEMI threats for information security: Remote command injection on modern smartphones. IEEE Trans. Electromagn. Compat. 57(6), 1752–1755 (2015)
10. Zhang, G., Yan, C., Ji, X., Zhang, T., Zhang, T., Xu, W.: DolphinAttack: Inaudible voice commands. In: Proceedings of the 2017 ACM SIGSAC Conference on Computer and Communications Security (CCS), pp. 103–117. ACM (2017)
11. Yan, Q., Liu, K., Zhou, Q., Guo, H., Zhang, N.: SurfingAttack: Interactive hidden attack on voice assistants using ultrasonic guided waves. In: Network and Distributed Systems Security (NDSS) Symposium (2020)
12. Vaidya, T., Yuankai, Z., et al.: Cocaine noodles: Exploiting the gap between human and machine speech recognition. In: 9th USENIX Workshop on Offensive Technologies (WOOT'15). USENIX Association (2015)
13. Carlini, N., Mishra, P., Vaidya, T., Zhang, Y., Sherr, M., Shields, C., Wagner, D., Zhou, W.: Hidden voice commands. In: 25th USENIX Security Symposium (USENIX Security 16), Austin, TX (2016)
14. Abdullah, H., Garcia, W., Peeters, C., Traynor, P., Butler, K.R.B., Wilson, J.: Practical hidden voice attacks against speech and speaker recognition systems. In: Network and Distributed Systems Security (NDSS) Symposium (2019)
15. Carlini, N., Wagner, D.: Audio adversarial examples: Targeted attacks on speech-to-text. In: 2018 IEEE Security and Privacy Workshops (SPW), pp. 1–7. IEEE (2018)
16. Schönherr, L., Kohls, K., Zeiler, S., Holz, T., Kolossa, D.: Adversarial attacks against automatic speech recognition systems via psychoacoustic hiding. In: Network and Distributed Systems Security (NDSS) Symposium (2019)
17. Qin, Y., Carlini, N., Cottrell, G., Goodfellow, I., Raffel, C.: Imperceptible, robust, and targeted adversarial examples for automatic speech recognition. In: International Conference on Machine Learning (ICML), pp. 5231–5240 (2019)
18. Yakura, H., Sakuma, J.: Robust audio adversarial example for a physical attack. Preprint (2018). arXiv:1810.11793

19. Chen, T., Shangguan, L., Li, Z., Jamieson, K.: Metamorph: Injecting inaudible commands into over-the-air voice controlled systems. In: Network and Distributed Systems Security (NDSS) Symposium (2020)
20. Yuan, X., Chen, Y., Wang, A., Chen, K., Zhang, S., Huang, H., Molloy, I.M.: All your Alexa are belong to us: A remote voice control attack against echo. In: 2018 IEEE Global Communications Conference (GLOBECOM), pp. 1–6. IEEE (2018)
21. Dong, Y., Liao, F., Pang, T., Su, H., Zhu, J., Hu, X., Li, J.: Boosting adversarial attacks with momentum. In: Proceedings of the IEEE Conference on Computer Vision and Pattern Recognition (CVPR), pp. 9185–9193 (2018)
22. Chen, S., Ren, K., Piao, S., Wang, C., Wang, Q., Weng, J., Su, L., Mohaisen, A.: You can hear but you cannot steal: Defending against voice impersonation attacks on smartphones. In: Distributed Computing Systems (ICDCS), 2017 IEEE 37th International Conference on, pp. 183–195. IEEE (2017)

A Survey on Secure Outsourced Deep Learning

Xu Ma, Xiaoyu Zhang, Changyu Dong, and Xiaofeng Chen

1 Introduction

The recent advances in deep learning [3] coupled with growth in computational capacities have improved the state of the art in many data-driven applications. For example, deep learning driven automatic detection and monitoring system can extract actionable information from vast amounts of data that in the past would have been impossible. Data analytics based on deep learning have transformed the practice in healthcare and financial systems.

However, deep learning algorithms require a large amount of storage and computation resources, and in general the performance of a machine learning model is highly influenced by the quality and volume of the training dataset. Therefore, the clients especially those with limited resources would prefer to outsource deep learning tasks to a cloud server who can provide almost unlimited storage and computation resources. But serious privacy and security issues also emerge from outsourced deep learning. Firstly, highly sensitive information can be included in the training data, such as healthcare and financial data. Directly uploading sensitive data in the clear to a potentially curious or malicious cloud brings high risks of information leakage. In addition, when distributed or collaborative learning has to be implemented over multiple datasets, privacy concerns also extend to malicious

X. Ma
State Key Laboratory of Integrated Service Networks (ISN), Xidian University, Xi'an, China

School of Software, Qufu Normal University, Qufu, China

X. Zhang · X. Chen (✉)
State Key Laboratory of Integrated Service Networks (ISN), Xidian University, Xi'an, China
e-mail: xfchen@xidian.edu.cn

C. Dong
School of Computing, Newcastle University, Newcastle upon Tyne, UK

© The Author(s), under exclusive license to Springer Nature Singapore Pte Ltd. 2021
X. Chen et al. (eds.), *Cyber Security Meets Machine Learning*,
https://doi.org/10.1007/978-981-33-6726-5_6

clients, who participate in the computation. Secondly, the cloud server has to be able to prove the correctness of the returned result and checking the proof should require a significantly less computational power than rerun the computation itself.

As described above, one of the most important security requirements of outsourced deep learning is privacy. Recall that a complete deep learning algorithm includes feature extraction, training, and prediction. Depending on which stage the outsourced computation occurs in, different privacy levels have to be satisfied. To summarize, privacy requirements in outsourced deep learning can be classified into data privacy and model privacy. Although in general privacy protection can be realized by encrypting or blinding the datasets or the models, this makes it difficult for the cloud server to run deep learning algorithms on the encrypted datasets. Therefore, how to achieve a tradeoff between privacy and utility in outsourced deep learning is the main objective.

Despite growing interest in the outsourced computation and deep learning domain, most of the existing contributions are scattered across different research areas. This article focuses on the crossover of the two research areas. We summarize techniques of outsourced computation that can be tailored to deep learning tasks, to achieve the best performance in various application scenarios. We then further review the most relevant literature and discuss the key pros and cons of various outsourced deep learning architectures. We conclude this paper by pinpointing future research directions.

Our Contributions The objective of this paper is to provide a comprehensive view on outsourced computation practices in deep learning. Specifically, we focus on the crossovers between outsourced computation and deep learning. Our main scope remains the outsourced deep learning domain, but for completeness we also review and discuss deep learning and outsourced computation. Overall, our contribution is threefold:

- We review the evolution of deep learning, including multilayer perceptron, recurrent neural network, convolutional neural network, and deep reinforcement learning; from these deep learning models extract the essential mathematical algorithms that can be realized by outsourced computation.
- We survey the high-significant articles related to outsourced computation, and discuss the cutting-edge outsourced techniques from the perspective of deep learning, pointing out the relation between outsourced computation and the fundamental mathematical operations extracted from deep learning algorithms.
- We particularly focus on outsourced deep learning, analyzing and comparing the underlying cryptographical techniques and outsourced architectures with respect to efficiency, security, and privacy; based on this analysis, we provide insights into the outsourced computation techniques selection strategies for outsourced deep learning.

Organization Deep learning and outsourced computation have been mostly researched independently. In recent years, the crossover between the two areas is increasingly emerged. We categorize these works into (1) overview of deep learning and its applications, (2) overview of the techniques of outsourced computation, and (3) reviews of the works at the intersection of deep learning and outsourced computation. As depicted in Fig. 1, we begin by reviewing the high-significant articles related to deep learning and introducing the background in Sect. 2. In Sect. 3, we introduced the preliminaries of outsourced computation. In Sect. 4, we focus on outsourced deep learning, highlighting the advantages of outsourced computation in addressing deep learning problems. We introduce and compare the state-of-the-art outsourced deep learning models, methods, and algorithms, aiming to help researcher in deep learning and outsourced computation in designing more efficient schemes. We then conclude this article with a view to future research directions in Sect. 5.

2 Deep Learning

In this section, we begin with a brief review of deep learning, and an introduction of the mainly used architecture is presented later. We then extract the underlying computation of deep learning in order to build a clear relationship between deep learning and outsourced computation.

2.1 Brief Survey on Deep Learning

Deep learning [37] is a sub-branch of machine learning, and it was originated from artificial neural network. Most of the early research on machine learning exploited shallow-structured architectures, such as Gaussian mixture models, support vector machines, and logistic regression. Generally, there is only one layer in these models responsible for transforming the input features into the output space. Although the shallow-architecture models are effective to solve many simple problems, their limited representation power causes difficulties when dealing with complicated application, such as speech and vision related application scenarios. These applications suggest the need of deep architecture to extract complex features and representations from the input. Deep Neural network is the most well-known deep learning model that has a sufficient number of hidden layers. Besides, other models with multiple layers can also be regarded as deep model, such as deep Gaussian process and deep random forests.

Table 1 lists the high-significant survey articles related to deep learning. Goodfellow [37] provided a tutorial book on deep learning which consists of the prerequisite knowledge, underlying optimization methods, and popular applications. In [69], Schmidhuber et al. summarized the relevant works and gave an encyclopedic survey on deep learning, including supervised and unsupervised learning, reinforcement learning, and provided short programs encoding deep and large networks. In [52],

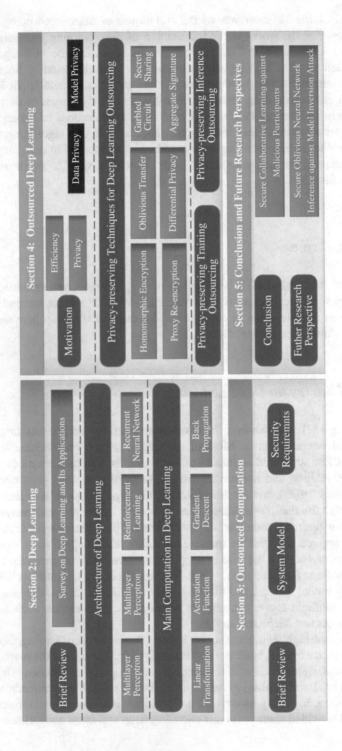

Fig. 1 The organization of this survey

Table 1 Summary of existing high-significant survey articles related to deep learning

Publication	One sentence contribution
[37]	A tutorial book on deep learning
[69]	An encyclopedic survey on deep learning
[52]	A comprehensive survey on deep learning
[38]	Applications of deep learning
[42]	
[47]	
[75]	
[48]	A survey on multilayer neural network and its applications
[5]	A comprehensive on deep reinforcement learning
[81]	A survey on multitask learning
[50]	Survey the use of deep learning for medical image analysis

Liu et al. presented the historical evolution of deep learning and its applications, summarized the underlying principles of several deep learning models. They also reviewed the applications of deep learning, such as speech recognition, computer vision, and signal processing [38, 42, 47, 75].

LeCun et al. [48] summarized the model, method, and algorithms of multilayer neural networks training and reviewed the application of Convolutions Neural Network (CNN) and Recurrent Neural Network (RNN) on image understanding and language processing. Arulkumanran et al. [5] surveyed reinforcement learning that covered central algorithms, including the deep Q-network, asynchronous advantage actor-critics, and trust region policy optimization. In [81], Zhang et al. gave a survey on multitask learning which is categorized into feature learning and selection approaches, low-rank approach, task clustering, and relation learning approach. Litjens et al. [50] surveyed the applications of deep learning in medical image classification, segmentation, object detection, and other tasks.

2.2 Architecture of Deep Learning

In this section, we briefly review the architectures of deep learning, including multi-layer perceptron, convolutional neural network (CNN), recurrent neural network (RNN), and reinforcement neural network. We only present a concept of these architectures, for more details please refer to [25, 37].

Multilayer Perceptron Multilayer perceptron (MLP) [25], which are also known as multilayer networks, refers to networks composed of multiple layers of percep-tron. MLP consists of three or more layers. Its simplest architecture consists of an input layer, a hidden layer, and an output layer. Each node of the network is a neuron that uses a nonlinear activation function and connects to every node in the following layer with a certain weight. Frequently used nonlinear activation functions such as sigmoid, tanh, ReLU, etc. are described in the following section. Supervised learning is used to optimize a MLP, by changing connection weights based on the loss minimization methods. The most well-known algorithm is backpropagation that is described in Sect 2.3.

Convolutional Neural Network CNN [37] is a regularized version of MLP and widely used in detection, segmentation, and recognition of objects and regions in images. Compared to the full connection of MLP, CNN takes advantage of the hierarchical pattern in data and assembles more complex patterns through convolution. Convolution is a special kind of linear transformation, and CNN is therefore defined as the neural network that use convolution in at least one of their layers. During the convolution stage, each unit is connected to local patches in the feature maps through a set of weights called a filter bank which is shared among all the units in a feature map. Typically, the architecture of CNN consists of convolutional layers, pooling layers, and nonlinear transformation layers, and several fully connected layers are used to achieve final classification.

Recurrent Neural Network RNN [37] is designed to take a series of input without a limited size. It remembers the past and the decisions are influenced by the experiences leaned from the past. During the training process, RNN remembers things learned from prior input while generating output. The output is determined not only by weights applied on inputs but also by the context based on prior inputs or outputs. So, the output of the same input could be different depending on previous inputs in the series. Because of its memorability, RNN is suitable for cases where the output depends on previous computations, like speech recognition, text recommendation, or DNA sequences analysis. Figure 2 shows a RNN with a hidden state.

Reinforcement Learning Reinforcement learning (RL) [73] is one of the methods of machine learning. It is about how software agent should take actions to maximize the cumulative reward in a particular environment which is typically described as

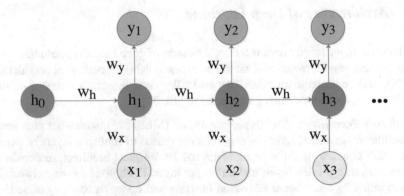

Fig. 2 An example of RNN with a hidden state

a form of Markov decision process. Compared to supervised learning, there is no training dataset in reinforcement learning. Therefore, in the absence of existing training data, RL learns from the experiences. During the learning process, the agent attempts its task through trial-and-error and rewards and punishment are used as signals for positive and negative behaviors. Also, it is different from unsupervised learning whose goal is to find the similarities between data points. The goal of RL is to find a suitable actions model to maximize the reward. Recent advances in RL have enabled a variety of applications such as games [73], robotics [39], and generative modeling [78].

2.3 Main Computation in Deep Learning

We introduce several mainly used computation in deep neural network. As shown in Fig. 3, the architecture of deep neural network consists of input layer, multiple hidden layers, and the output layer. The input layer is composed of features while the output layer stands for a classification. The layers are connected in a hierarchical manner, and the output of the previous layer is the input of the successive layer uses. Each node, which is also called a neuron, in the hidden layer and the output layer is associated with a coefficient vector and an activation function. The neuron firstly computes the weighted summation of its input, and then applies a nonlinear activation function to the summation (Fig. 4).

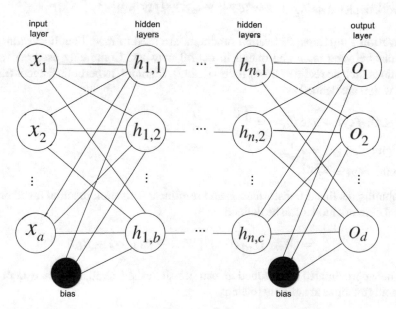

Fig. 3 An example of neural network with n hidden layers

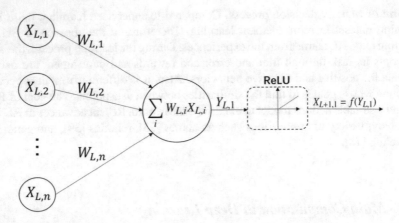

Fig. 4 Computations in one neuron

Linear Transformations Linear transformation is the commonest operation in neural network, which is composed of matrix additions and multiplications, can be described as:

$$Y = WX + B,$$

where W denotes the weight matrix, X denotes input vector, Y denotes the output vector, and B denotes the bias vector, for each layer. For the first layer of the neural network depicted in Fig. 3, $X \in \mathrm{R}^{a \times 1}$, $Y \in \mathrm{R}^{b \times 1}$, $W \in \mathrm{R}^{b \times a}$.

Activation Functions Activation functions are used to model nonlinear transformations between input and output in neural network. Commonly used activation functions include rectified linear units (ReLU), Sigmoid, hyperbolic tangent (tanh), etc., which are defined as:

- ReLU: $f(y) = \max(0, y) = \begin{cases} 0 & y \leq 0 \\ y & y > 0 \end{cases}$.
- Sigmoid: $f(y) = \frac{1}{1+e^{-y}}$.
- tanh: $f(y) = \frac{e^{2y}-1}{e^{2y}+1}$.

Combining the linear transformation and nonlinear activation function together, the neural network model can be defined as:

$$Y := (W_L \cdot f_{L-1}(\cdots f_1(W_1 X + B_1) \cdots) + B_L).$$

The approximated tanh function can be described similarly, since tanh and sigmoid functions are closely related:

$$tanh(y) = 2 \cdot sigmoid(2y) - 1.$$

Back Propagation Algorithm Parameters optimization of a neural network is a nonlinear optimization problem. Gradient descent [6, 29, 68] is one of the mostly used algorithms which is based on the observation that a multi-variant function $F(x)$ decreased fastest in the negative direction of its gradient $\nabla F(a)$ if the function is defined and differentiable in the neighborhood of a point a. Generally, gradient descent (1) starts at a randomly selected point, and then (2) the learning algorithm computes the gradient of the function being optimized, and (3) updates the parameters using the gradient so as to optimize the objective function. This process will not stop until the algorithm converges to a local optimum.

Backpropagation [68] is a widely used algorithm which is based on gradient descent in feedforward neural network for supervised learning. Backpropagation computes the gradient of the loss function, which is defined as the total error between the output of the neural network and the objective value, with respect to the parameters (weights) of the neural network by the chain rule. After each forward pass through the network, backpropagation performs a backward pass by adjusting the network's weight parameters. Let $\mathbf{w}$ be the flattened weight vector of the neural network and E be the loss function. The partial derivative is computed as $\frac{\partial E}{\partial w_j}$ of E with respect to each parameter in $\mathbf{w}$. In each iteration, the weights can be updated as: $w_j = w_j - \alpha \frac{\partial E}{\partial w_j}$, where α is the learning rate. The iteration process will continue until the output of loss function satisfies a predefined accuracy trapdoor.

3 Outsourced Computation

In this section, we first review the evolution of outsourced computation, and then propose the system model and security requirements of outsourced computation.

3.1 Brief Survey on Outsourced Computation

Outsourced computation has been studied in recent years by a plenty of researchers due to the rise of cloud computing. Outsourced computation protocols enable a computationally weak client to securely outsource some computation to a powerful server. Generally speaking, outsourced computation can be classified into two categories: universal-oriented solutions and function-specific solutions. Function-specific solutions focus on a specific class of functions, while the universal solutions can be applied to a wide range of computable functions. However, function-specific outsourced computation protocols provide higher efficiency compared to universal-orient outsourced protocols. For example, the universal solutions provided in [19, 67] rely on probabilistically checkable proofs [10], fully homomorphic encryption [35], and homomorphic authenticators [33] which are hard to be utilized in practice due to the extremely high computation and storage costs. A summary

Table 2 Summary of existing high-significant articles related to outsourced computation

Publication	One sentence contribution
[71]	A comprehensive survey on outsourced computation
[79]	A comprehensive survey on verifiable computation delegation
[58]	Secret computations using insecure auxiliary devices
[45]	Efficient arguments
[60]	Computationally sound proof
[8]	Delegatable homomorphic encryption &applications to outsourced computation
[11]	Verifiable computation for high-degree polynomial functions
[7]	Outsourced computation on outsourced dataset
[67]	A primitive of publicly verifiable computation delegation
[74]	Publicly verifiable computation delegation for polynomial evaluation
[19]	Practical homomorphic MACs for arithmetic circuits
[71, 79]	Survey of outsourced computation

of existing high-significant articles related to outsourced computation is listed in Table 2.

The problem of securely outsourcing expensive computation has been well-studied in the cryptography community. One of the main security issues of outsourced computation is how can the clients check the correctness of the results provide by the cloud without much computational efforts. Efficient arguments [45] and computationally sound proofs [60] are considered in previous work to realize verifiability in outsourced computation.

Benabbas et al. [11] proposed verifiable delegation of computation and presented the first practical outsourced scheme for high-degree polynomial functions over large dataset. In [8], Barbosa et al. proposed a new privacy protection cryptographical tool called delegatable homomorphic encryption. A client can thus delegate the evaluation of circuits to untrusted servers over encrypted data. Backes et al. [7] proposed a method to solve the problem of outsourced computation over outsourced dataset. In this application scenario, the client firstly outsources its data to a server which will later be required to compute a function over the outsourced data. The scheme is based on homomorphic message authenticator to realize efficient verification. However, the scheme can only be used for quadratic polynomials evaluation.

In [67], Parno et al. extended the definition of outsourced computation delegation and proposed public verifiable outsourced computation primitive. This primitive can be used to construct publicly verifiable outsourced computation scheme. In [74], Song et al. investigated publicly verifiable outsourced computation for polynomial evaluation with the support of multiple data sources. In [19] , Dario et al. proposed a homomorphic message authenticator that allows the holder of an evaluation key to perform computations over previously authenticated data, in such a way that the produced tag can be used to certify the authenticity of the computation.

In [71], Shan et al. summarized the cutting-edge techniques for specific secure outsourced computation, including matrix computation, mathematical optimization, etc. In [79], Yu et al. surveyed the existing research works on verifiability computation, comparing and discussing the pros and cons according to performance and security requirements.

3.2 System Model

In [34], Gennaro et al. provided the formal definition of outsourced computation. Specifically, an outsourced computation scheme consists of the four algorithms defined below.

- **KeyGen**$(F, \lambda) \rightarrow (PK, SK)$: Based on the security parameter λ, the randomized key generation algorithm generates a public key that encodes the target function F, which is used by the worker to compute F. It also computes a matching secret key, which is kept private by the client.
- **ProbGen**$_{SK}(x) \rightarrow (\sigma_x, \tau_x)$: The problem generation algorithm uses the secret key SK to encode the function input x as a public value σ_x which is given to the worker to compute with, and a secret value τ_x which is kept private by the client.
- **Compute**$_{PK}(\sigma_x) \rightarrow \sigma_y$: Using the client's public key and the encoded input, the worker computes an encoded version of the function's output $y = F(x)$.
- **Verify**$_{SK}(\tau_x, \sigma_y) \rightarrow y \cup \bot$: Using the secret key SK and the secret "decoding" τ_x, the verification algorithm converts the worker's encoded output into the output of the function, e.g., $y = F(x)$ or outputs $\bot$ indicating that σ_y does not represent the valid output of F on x.

An outsourced computation delegation scheme is correct if the problem generation algorithm produces values that allow an honest worker to compute values that will verify successfully and correspond to the evaluation of F on those inputs.

3.3 Security Requirements

In this section, we review the formal security definition of outsourced computation, for more details please refer to [21, 67].

The security requirement of privacy is for the input/output of the computation task. The formal definition of privacy can be described in the sense of an indistinguishability argument.

Definition 1 (Privacy) An outsourced protocol that achieves the privacy for the input/output of F is for any probabilistic polynomial time adversary A,

$$Adv_A(F, \lambda) \leq \text{negl}(\lambda),$$

where $Adv_A(F, \lambda) = |\Pr[b' = b] - \frac{1}{2}|$ is defined as the advantage of A is the experiment as follows:

Experiment $\mathbf{Exp}_A[F, \lambda]$
$(PK, SK) \leftarrow \mathsf{KeyGen}(F, \lambda)$
$(x_0, x_1) \leftarrow A^{\mathsf{PubProbGen}(\cdot)}(pk)$
$(\sigma_0, \tau_0) \leftarrow \mathsf{ProbGen}_{sk}(x_0)$
$(\sigma_1, \tau_1) \leftarrow \mathsf{ProbGen}_{sk}(x_1)$
$b \leftarrow_R \{0, 1\}$
$b' \leftarrow A^{\mathsf{PubProbGen}_{sk}(\cdot)}(PK, x_0, x_1, \sigma_b)$
If $b' = b$, output 1, else 0.

The adversary A is allowed to request the encoding of any input he desires in the experiment above. **PubProbGen**$(\cdot)$ calls on **ProbGen**(x) and returns only the public part σ_x.

Verifiability An outsourced computation scheme is secure if a malicious worker cannot persuade the verification algorithm to accept an incorrect output. In other words, for a given function F and input x, a malicious worker should not be able to convince the verification algorithm to output $\hat{y}$ such that $F(x) \neq \hat{y}$.

Efficiency An outsourced computation delegation scheme can be outsourced if it permits efficient generation and efficient verification. This implies that for any x and any σ_y, the time required for problem generation plus the time required for verification is $O(T)$, where T is the time required to compute $F(x)$.

Besides the essential security requirements described above, there are some additional properties which are expected in some specific scenarios, such as public verifiability, non-interactiveness, and homomorphic authentication.

Non-interactiveness In [34], Gennaro et al. proposed non-interactive outsourced computation protocol that allows the worker (server) to return a computationally sound, non-interactive proof that can be verified in $O(m)$ time, where m is the output length of the output function F. In this protocol, the client sends a single message to the worker and vice versa. The crucial efficiency requirement is that input preparation and output verification must take less time than computing F from scratch.

4 Outsourced Deep Learning

In this section, we review the related works and give a comprehensive analysis on outsourced deep learning. We first introduce the motivation and privacy concerns in outsourced deep learning. Secondly, we present the privacy-preserving techniques, including encryption, signature, secret sharing, etc. We classify the existing solutions into privacy-preserving training and privacy-preserving inference, and then summarize, compare, and discuss these works with respect to security level and efficiency performance.

4.1 Brief Review on Outsourced Deep Learning

In Table 3, we list the high-significant articles related to outsourced deep learning. In this section, we will specify the techniques used in these works, and present a detailed comparison among these works in terms of privacy and efficiency performance.

Outsourced deep learning is defined as the training or prediction tasks of deep learning that are completed with the help of the servers without leaking private information about the data. It can be classified into two outsourced training and outsourced inference. Outsourced training algorithms can be used for outsourced inference, but not vice versa. Because inference is actually a forward pass through of the learning model, which is already included in the training process. Shokri et al. [72] proposed a secure multiparty deep learning based on the technique of parameter sharing. To solve the privacy leakage problem in [72], Anno et al. [4] introduced privacy-preserving deep learning using additively homomorphic encryption.

Table 3 Classification of existing privacy-preserving deep learning outsourced schemes from the system model

	Classification	Publications & Main techniques
Individual learning outsourcing	One server	[82]-ΓΗΕ+TSE [63]-FHE+SGD [54]-FHE+FHECS
Collaborative learning	One server	[80]-DHE [72]-PPS [40]-DP [64]-SGX [66]-PATE+GAN [13]-SS+AE [49]-MKHE [4]-AHE [55]-AHE+AS [41]-AHE+DP [57]-PRE+AS
	Multiple non-colluding servers	[2, 62]-OT+GC+SS [61]-SS+OT+3PC [20]-MPC+SS+DP
Secure inference outsourcing	One server	[43]-SMM+HE [22]-MKHE
	Multiple non-colluding servers	[56]-AHE+AS

Mohassel et al. [62] presented an outsourced deep learning protocol in two non-colluding servers model. The neural network model is additively shared to the servers. In prediction phase, the client also splits its input into two parts which are sent to the servers, respectively. However, the protocol is interactive, the client has to stay online during the learning process and the security model is honest-but-curious. In [55], Ma et al. proposed a secure multiparty aggregation scheme with verifiability in malicious adversary, and used it to construct a privacy-preserving multiparty learning algorithm. Zhang et al. [83] proposed an outsourced batch matrix multiplication scheme which was extended to privacy-preserving one-layer perceptron deep learning.

Federated learning [46] proposed by Google AI can also be viewed as a kind of outsourced deep learning where the parameters are exchanged and aggregated by a cloud server. Compared to traditional collaborative learning on distributed dataset where all the local datasets are uploaded to a central server, federated learning exhibited many advantages. The general principle of federated learning is training on local dataset and exchanging the parameters originated from each local learning model. Since the local data never leaves its domain, federated learning addresses the security issues such as data privacy and data access rights. However, as pointed out by Fredrikson [30], federated learning is vulnerable against model inversion attack which enables the adversary to extract private information about the training data using white box attack to the learned model.

4.2   Privacy Concerns in Outsourced Deep Learning

The great advancement of deep learning in recent years is largely due to the improvement of computation power and storage capability. For those who has limited computation and storage resources, it is difficult for them to run deep learning algorithms independently. Therefore, outsourced deep learning tasks to a powerful infrastructure such as a cloud provides an alternative choice. On the other hand, the accuracy of a learning model is highly influenced by how much data used for training. Generally, multiple parties would benefit from collaborative learning where the machine learning algorithm runs on the aggregated dataset. A trivial method for multiparty learning is outsourced the data from multiple parties to a server to construct a large enough dataset center and process them centrally.

However, privacy concerns on the training data and the learning model are raised when we outsource deep learning tasks to an untrusted server. **Data privacy** denotes that the data used for training must be kept private against the server during the outsourced deep learning process. In some of the application scenarios, such as the financial or healthcare systems in which the data is highly sensitive, it is prohibited for these organizations to share its data with any other third party. **Model privacy** refers to that the trained model has to be protect. The learning model is abstracted from the training data, and it is a valuable property that should be protected. Model inversion attack was firstly introduced in [30] and extended in

[31]. In model inversion attack, adversarial access to the trained model is abused to learn sensitive information about individuals, such as reconstructing recognizable images of people's face.

To tackle the privacy concerns in outsourced deep learning, various approaches have been proposed. In [80], Yuan et al. propose a secure multiparty collaborative learning in cloud computing. In the proposed scheme, each party encrypts his private data using BGN homomorphic encryption and uploads the encrypted data to the server. The cloud server can execute most of the operations of the learning algorithm over the ciphertexts and return encrypted results to the client. However, since BGN is not a fully homomorphic encryption scheme [17], each participant has to decrypt and re-encrypt the intermediate values during the learning process. Thus, the interactiveness between the server and each participant brings high communication overhead, and the decryption and encryption computation costs for every in the learning process make the scheme difficult to be used in real application scenarios.

In [9], Barni et al. propose a privacy-preserving neural network prediction scheme, in which a user asks a services provider to run a neural network prediction on protected personal data. Meanwhile, model privacy, including the weight parameters, activation functions, and neural network topology, is protected against the user. The scheme relies on secure multiparty computation techniques, such as private scalar product and private polynomial evaluation, and it is proved secure under semi-honest security model. In the following sections, we will classify the related privacy-preserving articles and give a comprehensive analysis with respect to the underlying techniques, security model, efficiency, and performance improvements.

4.3 Privacy-Preserving Techniques for Outsourced Deep Learning

Homomorphic Encryption An additively homomorphic encryption scheme satisfies the following properties: given two ciphertexts $c_1 := E(pk, x_1)$ and $c_2 := E(pk, x_2)$, there is a public key operation $\oplus$ such that $E(pk, x_1 + x_2) = c_1 \oplus c_2$. Examples of such schemes include ElGamal encryption [28], Paillier's encryption [65], and DGK encryption [24]. An example of multiplicative homomorphic encryption is presented as follows: given the ciphertexts $C, \hat{C}$ of m and $\hat{m}$, respectively.

$$C = (c_1, c_2) = (g^{k_1}, y^{k_1} m);$$

$$\hat{C} = (\hat{c}_1, \hat{c}_2) = (g^{k_2}, y^{k_2} \hat{m}).$$

The ciphertext $\bar{C}$ of $m\hat{m}$ can be computed as $C\hat{C}$. More specifically:

$$\bar{c}_1 = c_1 \hat{c}_1 = g^{k_1 + k_2}, \bar{c}_2 = c_2 \hat{c}_2 = y^{k_1 + k_2} m\hat{m}.$$

If the plaintext space is small, it is easy to transfer this multiplicative homomorphic encryption to additively homomorphic encryption by lifting the message m to exponentiation g^m, such that the ciphertext is

$$C = (c_1, c_2) = (g^{k_1}, y^{k_1} g^m).$$

Secret Sharing Secret sharing [70] refers the methods of distributing a secret among a group of participants. Each participant of the group is allocated a share of the secret, and the secret can be reconstructed by a subset of the group. The (n, t)-threshold secret sharing is defined as in a group of n participants, in which each participant is given a share. The secret can be reconstructed in such a way that any $t\,(t \leq n)$ or more participants can work together to reconstruct the secret. One of the most widely used secret sharing is additively secret sharing, in which the secret is shared between two or more parties such as that the addition of all shares yields the secret. The operations are usually performed in the ring $\mathbb{Z}_{2^l}$, and each share is represented as an l-bit integer. Take an example in two-party setting, the secret s can be secretly shared as follows: (1) randomly select a number $r \in_R \mathbb{Z}_{2^l}$; (2) compute $s - r \mod 2^l$; (3) set the shares as $\langle s \rangle_0 = r$ and $\langle s \rangle_1 = s - r \mod 2^l$. The secret s can therefore be reconstructed as $s = \langle s \rangle_0 + \langle s \rangle_1 \mod 2^l$.

Oblivious Transfer Oblivious transfer (OT) [36] is a fundamental cryptographical primitive in secure multiparty computation. The OT protocol involves a sender S and a receiver R. R selects and receives a message from a set of messages owned by S who knows nothing about which message was selected, and R should also know nothing about the unselected message. A commonly used form of oblivious transfer is 1-out-of-2 OT, in which the sender S has two 1-bit messages x_0 and x_1 and the receiver R has a bit b indicating the index of the desired message. After performing the protocol, R obtains x_b without learning anything about x_{1-b} or revealing anything to S

Garbled Circuits Garbled circuits, introduced in [77], are an efficient method for generic secure two-party computation in which two mutually untrusted parties jointly compute a function $f(a, b)$ without leaking any information about their private inputs a, b except what is leaked by the final output. Briefly speaking, the function to be computed is described as a Boolean circuit with two-input gates AND and XOR. Then one of the parties (Alice) garbles the circuit and sends it to the other party (Bob). Alice generates and assigns two k-bit random numbers for each wire in the circuit, one stands for Boolean semantic 0 and one for 1, where k is the security parameter. Then, Alice garbles the circuit according to the truth table by replacing 0 and 1 with the random values, and sends the garbled table to Bob after permutation. The garbling process can be done with symmetric encryption, such as AES. Meanwhile, Alice sends the random numbers corresponding to her input to Bob who also receives the random numbers corresponding to his input from Alice through OT. Then, Bob goes through all gates one by one and continues the evaluation until he reaches the output. Finally, Bob reveals the random numbers to Alice who can map them back to Boolean values.

Proxy Re-encryption In [12], Blaze et al. introduced the concept of PRE (Proxy Re-encryption). Using PRE, one can convert the ciphertexts for one key into ciphertexts for another with a proxy function and a public proxy key. PRE has the following properties:

- **Bidirectional:** Encryption delegation from user A to user B allows re-encryption in both directions, i.e., from A to B and vice versa ($B \Leftrightarrow A$).
- **Collusion nonresistance:** Collusion of user A and proxy can recover user B's secret key. The PRE scheme in [12] is secure under the assumption that there is no collusion between proxy and users.
- **Homomorphic:** Note that the PRE is ElGamal-based, so it is multiplicatively homomorphic over the ciphertexts. Upon the input ciphertexts C of m and $\hat{C}$ of $\hat{m}$,

$$C = (c_1, c_2) = (m \cdot g^{r_1}, g^{s_a r_1}), \hat{C} = (\hat{c}_1, \hat{c}_2) = (\hat{m} \cdot g^{r_2}, g^{s_a r_2}).$$

The encryption $\bar{C}$ for $m \cdot \hat{m}$ can be computed as $C \otimes \hat{C}$. More specifically,

$$\bar{C} = C \otimes \hat{C} = (c_1 \cdot \hat{c}_1, c_2 \cdot \hat{c}_2) = (m \cdot \hat{m} \cdot g^{r_1 + r_2}, g^{s_a(r_1 + r_2)}).$$

Bilinear Aggregate Signature In many applications, the verifier must verify many signatures on different messages generated by different signers, such as the verification of a chain of n certificates signed by n Certificate Authorities (CAs). To improve the verification efficiency, the aggregate signature was proposed in [14]. Suppose that each user u_i owns a key pair (pk_i, sk_i), where σ_i is the signature on message m_i signed by u_i. Then, there is a public aggregation algorithm that generates a short aggregate signature σ from $\sigma_1, \sigma_2, \ldots, \sigma_n$. Additionally, there is an aggregate verification algorithm that takes $pk_1, pk_2, \ldots, pk_n, \sigma$ and all messages $m_1, m_2, \ldots, m_n$ as input and determines the correctness of the aggregate signature.

Differential Privacy Differential privacy [26, 27] constitutes a strong standard for privacy guarantee. It guarantees that the outcome of any analysis are not substantially affected by the removal or addition of a single data item. In other words, the goal of differential privacy is to reveal only results such that it is impossible to tell if a certain record was, or was not, in the dataset that the function was computed on.

A randomized function $\mathcal{K}$ fulfills ϵ-differential privacy iff all datasets D and D' differing on at most one tuple, and for all $S \subseteq Range(\mathcal{K})$,

$$\Pr[\mathcal{K}(D) \in S] \leq exp(\epsilon) \times \Pr[\mathcal{K}(D') \in S]$$

, where ϵ stands for privacy budget that controls the amount of difference induced by two neighboring datasets. In [26], Dwork et al. presented a general method to preserve differential privacy for any function f using Laplace mechanism. The mechanism exploits the global sensitivity of function f over any two neighboring datasets, and injects appropriately chosen random noise to the answer of the query function f on the database. For more details, please refer to [26].

4.4 Taxonomy Standard

Differentiated by the different phase of outsourced deep learning, we classified outsourced deep learning into (1) privacy-preserving training outsourcing and (2) privacy-preserving inference outsourcing. Note that the techniques utilized in collaborative learning can be trivially extended into individual learning outsourcing. Individual learning outsourcing can be viewed as a special case of collaborative learning where there is only one participant. Considering the system model, outsourced deep learning algorithms can be further classified into individual learning and collaborative learning. For each category, there might be one or more servers. We describe the classification of existing privacy-preserving deep learning outsourced schemes in Table 3. All the abbreviations in Table 3 are listed in Table 4.

Individual learning is mostly used for one client with limited computation or storage resources that outsources the deep learning task to one cloud server. Usually, homomorphic encryption [35] is used to encrypt the original dataset, while allowing the cloud server to implement deep learning algorithms on the encrypted data without having to decrypt it. However, using homomorphic encryption to process the whole dataset and running deep learning algorithms on the encrypted dataset incur high latency for outsourced deep learning.

It has been shown that in neural network learning significant accuracy improvements can be achieved by incorporating more data into the training set. However, collecting data from distributed datasets may be challenging when privacy is a major concern. A plenty of research work have been done towards the privacy-preserving collaborative learning. Secure multiparty computation, secret sharing, homomorphic encryption, and differential privacy are commonly used cryptographical tools in privacy-preserving collaborative learning, each of which has its own pros and cons. In the following sections, we will present a comparison on those cryptographical techniques based outsourced deep learning algorithms.

4.5 Privacy-Preserving Training Outsourcing

We describe in this section the solutions for privacy-preserving training outsourcing. As presented in Table 3, privacy-preserving training outsourcing can be further classified into individual learning outsourcing and collaborative learning outsourcing. Further, the related works can be classified into one server model and multiple (two) non-colluding model, based on how many servers are involved in the scheme. We therefore begin by reviewing the main techniques used in individual learning

Table 4 List of the abbreviations in alphabetic order

Acronym	Explanation
3PC	3-Party computation
AHE	Additively homomorphic encryption
AE	Authenticated encryption
AS	Aggregation signature
DHE	Doubly homomorphic encryption
DP	Differential privacy
FHE	Fully homomorphic encryption
FHECS	Fully homomorphic encryption combing and switching
GAN	Generative adversarial network
GC	Garbled circuit
HE	Homomorphic encryption
MKHE	Multi-key homomorphic encryption
MPC	Multiparty computation
OT	Oblivious transfer
PATE	Private aggregation of teacher ensembles
PPS	Partially parameter sharing
PRE	Proxy re-encryption
RG	Repeated Gompertz
RP	Random projection
PAHE	Packed additively homomorphic encryption
SMC	Secure multiparty computation
SGX	Secure guard extensions
SGD	Stochastic gradient decent
SS	Secret sharing
SMM	Secure matrix multiplication
TSE	Taylor series expansion
USC	Unconditionally secure comparison

and collaborative learning, then summarize and compare the performance of these schemes with respect to adversary model, privacy protection, and effectiveness.

Individual Learning Outsourcing In [82], Zhang et al. proposed an outsourced deep learning scheme based on FHE [17]. The learning efficiency is improved by outsourcing most of the computations to a semi-honest cloud server. To protect data and model privacy, training data and the initialized neural network are encrypted using FHE before being uploaded to the cloud server. Moreover, the nonlinear activation function (sigmoid function) is approximated by polynomials involving only additions and multiplications using Taylor theorem. Based on the homomorphism of FHE, the cloud server can run backpropagation algorithm without decrypting the ciphertext and return an encrypted model to the data owner. The security proof shows that data privacy and model privacy are satisfied. However, for each iteration of the learning process, the client has to decrypt and re-encrypt the cyphertexts returned from the cloud server, which makes the scheme fully interactive and incurs high computation and communication overhead to the data owner.

To realize a completely non-interactive and privacy-preserving individual learning, Nandakumar et al. [63] proposed a fully homomorphic encryption based stochastic gradient decent algorithm. The proposed scheme introduces a few tricks, such as neural network simplification, appropriate selection of data representation and resolution, and ciphertext packing techniques. Moreover, the neural network is trained in the fix-point domain and achieves convergence with reasonable accuracy. In the proposed scheme, the data owner encrypts the training data and initialized neural network model and uploads these ciphertexts to the cloud server. After that, the data owner can go offline until the cloud server returns the learned model. Although, the scheme achieves an overall classification accuracy approximately 97.8%, it is seriously limited by its long training latency. The experiments results showed that it takes about two hours with mini-batch including 60 samples on a machine that has two Intel Xeon E5-2698 16-cores CPUs (running at 2.30 GHz) and 250 GB RAM.

Recently, Lou et al. [54] proposed Glyph, which enables individual private deep learning over encrypted data. The scheme also relies on fully homomorphic encryption, but the efficiency and accuracy are improved by switching between TFHE [23] and BGV cryptosystems [17]. The correctness of such transformation originated from a recent work [16], which demonstrates the feasibility of combining and switching between TFHE and BGV via homomorphic operations. The nonlinear activations are implemented by logic-operation-friendly TFHE, while multiply-accumulation operations are performed with vectorial-arithmetic-friendly BGV. Moreover, transfer learning is incorporated into the scheme to reduce the number of trainable layers so that the training latency can be further reduced. The experimental results show that Glyph reduces the training latency by 99% over the prior work in [63] on various encrypted datasets (Table 5).

Table 5 Comparison of individual learning outsourced schemes

Publication	Main concept	Interactiveness
Zhang et al. [82]	–Encrypts the input data using the BGV encryption [17] –Approximate the Sigmoid function as a polynomial function	Interactive
Nandakumar et al. [63]	–Encrypts input data and model with FHE –Network simplification and ciphertext packing	Non-interactive
Lou et al. [54]	–Fully homomorphic cryptosystem transformation –Knowledge transfer	Non-interactive

The security models of these schemes are honest-but-curious
All the schemes have accuracy loss which depends on the activation function approximation method

Collaborative Learning (One Server Model) Learning from data collected by multiple parties has many potential applications. However, collecting massive data from distributed datasets presents new privacy issues and new solutions which enable multiple parties to jointly train a global learning model without revealing their private data are urgently demanded. Many works have been done towards privacy-preserving collaborative learning. In Table 6, we present the comparisons of the related schemes.

In [80], Yuan et al. proposed neural network learning in cloud computing, which enables two-party learning over arbitrarily partitioned data. The scheme is based on a special homomorphic encryption proposed by Boneh et al. [15], which supports one multiplication and unlimited number of addition operations. Sigmoid is approximated by a polynomial using Maclaurin series expansion method, but this approximation introduces an accuracy loss of the learned model. Moreover, the scheme is fully interactive since the participants have to decrypt the cyphertexts and re-encrypt them during each round of the learning process. Therefore, the computation and communication overhead are linearly related with the complexity of the neural network, database size, and number of participants.

Shokri et al. [72] considered a different way to realize privacy-preserving deep learning in multiparty scenarios. They abandoned cryptographical encryptions to process the data or learning model and design an efficient learning algorithm from partially parameter sharing which stimulate the following research on federated learning. A distributed selective stochastic gradient decent protocol was proposed in this paper. The participants train independently on their own datasets and then selectively share small subsets of parameters with other participants. The sharing process is implemented asynchronously, and each participant fully controls which gradient to share and how often. This method reaches a tradeoff between utility and privacy protection. The experiment results show that the learning accuracy of the proposed scheme is higher than standalone learning even when only 1 percent of parameters are shared.

Phong et al. [4] pointed out that in the system proposed by Shokri et al. [72], even a small portion of gradient parameters shared and stored on the cloud server can be exploited. They analyzed the information leakage problem about the training data through several computation examples related to the training process, and the results showed that the true value of the training data can be guessed with a non-negligible probability. In [72], Shokri et al. showed how to protect gradient privacy with differential privacy. However, a differential privacy protection mechanism harmed learning accuracy which is the main appeal of deep learning. To protect the gradient privacy without sacrificing learning accuracy, Phong et al. [4] proposed a new outsourced deep learning method by encrypting the gradients using additively homomorphic encryption before uploading to the cloud server. The scheme leaks no information about participants' gradient but also incurs increased computation and communication costs. Moreover, all the participants that share the same secret key of the homomorphic encryption are vulnerable against collusion attacks.

Table 6 Comparison of collaborative learning outsourced schemes with one server

Publication	Main concept	Interactiveness	Security model
Yuan et al. [80]	– Supports two data owners. – Uses a special homomorphic encryption to encrypt the data and model parameters. – Maclaurin series expansion	Interactive	Honest-but-curious
Shokri et al. [72]	– Gradient sharing. – Asynchronous stochastic gradient decent	Interactive	Honest-but-curious
Phong et al. [4]	– Using additively homomorphic to encrypt local gradients – Asynchronous stochastic gradient decent	Interactive	Honest-but-curious
Li et al. [49]	– Using multi-key fully homomorphic to encrypt the data and model parameters – Stochastic gradient decent	Non-interactive	Honest-but-curious
Ma et al. [55]	– Using additively homomorphic to encrypt the local gradients. – Using aggregate signature to verify the result. – Asynchronous stochastic gradient decent	Interactive	Malicious
Hao et al. [41]	– Using symmetric homomorphic to encrypt the local gradients. – Using differential privacy technique to perturb local gradients. – Asynchronous stochastic gradient decent	Interactive	Honest-but-curious Collusion resistant
Hamm et al. [40]	– Using knowledge transfer to train a student mode to train a student model. – Using technique to perturb teacher model's classification.	Non-interactive	Fully trusted server
Papernot et al. [66]	– Using knowledge transfer and differential privacy to train a student model. – Semi-supervised learning with GAN.	Non-interactive	Honest-but-curious
Ma et al. [57]	– Using knowledge transfer to train the model. – Using proxy re-encryption to encrypt local model's classification	Non-interactive	Malicious
Bonawitz et al. [13]	– Using secret sharing and double masking to blind the local gradients	Interactive	Malicious
Ohrimenko et al. [64]	– Using SGX to store secret data and execute secure computations	Non-interactive	Malicious

A method to address this problem was proposed in [49]. The basic idea is to encrypt the gradient with multi-key fully homomorphic encryption [53]. The advantage of multi-key homomorphic encryption lies in that it is unnecessary to share one secret key among all the participants. Instead, each participant owns a pair of public/secret keys. The local gradients are encrypted with different keys and uploaded to the cloud server and all the ciphertexts can still be aggregated together due to multi-key homomorphism. On receiving the encrypted result, all the participants work together through secure multiparty computation to decrypt it. Taking account of the high computation cost of multi-key fully homomorphic encryption and the complexity of deep learning, the scheme is hardly applicable in real application scenarios.

The related works mentioned above are secure in the honest-but-curious model (semi-honest model or passive model), where the cloud server and all the participants will honestly follow the protocol as prescribed but try to learn extra private information from the messages seen during the protocol execution by more computation. The honest-but-curious security model enables the development of efficient protocols, but it is also a weak security model. In [55], Ma et al. proposed a new method for secure collaborative learning under the malicious adversary model where the adversary is assumed to be able to deviate from the prescribed protocol at any time. In the proposed scheme, an aggregate signature of the gradient is also generated when a participant uploads the encrypted local gradients to the cloud server. The aggregate signature can later be aggregated into an aggregate signature of the aggregated gradients so that the participants can verify its correctness by checking whether the aggregated gradients and the signature form a valid pair of a message and its signature. Experiments demonstrate that the proposed scheme enables significantly efficient training compared to training on the encrypted dataset. However, the scheme is secure under the non-colluding assumption.

Hao et al. [41] proposed a federated learning scheme that is secure against collusion attack. Symmetric homomorphic encryption and differential privacy techniques are combined to protect the local gradients. Moreover, the scheme is based on stochastic gradient decent method so that it is tolerable of arbitrary subset of user dropping out with negligible accuracy loss. Specifically, for each epoch of the training process, each user computes the local gradient G_μ and perturbs it with Laplace noise $Lap(\frac{\Delta f}{\epsilon})$. Then, the encrypted and blinded gradient $C_\mu = Enc_{sk}(G_\mu + Lap(\frac{\Delta f}{\epsilon}))$ is sent to the server who executes the aggregation operation by $C_{add} = C_1 + C_2 + \cdots + C_n = Enc_{sk}(\sum_{\mu=1}^{n} G_\mu)$. The noises are almost eliminated from the summation due to the symmetric distribution of the Laplace mechanism. However, the scheme is secure under the honest-but-curious model.

Hamm et al. [40] proposed a secure multiparty learning scheme based on knowledge transfer [32] and differential privacy. In particular, there is a trust authority in the framework who is responsible for collecting the locally trained classifiers (teacher model). These classifiers are not aggregated directly to generate the global classifier but used for generating labels for non-sensitive, auxiliary, and unlabeled data. When generating labels for the auxiliary data, the authors proposed a new aggregation method, in which each sample is weighted by the confidence of the ensemble. The labeled data constitutes a new training set for the student model which will be perturbed with ϵ-differential privacy before releasing. Compared to a non-private solution, its generalization error converges with a fast rate of $O(\epsilon^{-2}M^{-2})$ and $O(N^{-1})$, where N is the number of unlabeled auxiliary data and M is the number of parties. However, there is a serious weakness in [40], that is, the local classifiers are collected by a fully trusted server. This assumption might not be practical in the real application scenarios. As pointed out by Fredrikson et al. in [31], the prediction or the internal parameters of a model can be used to infer sensitive information about the training data.

Papernot et al. proposed a strengthened scheme in [66] to address the privacy leakage problem in [40]. They proposed PATE, for Privacy Aggregation of Teacher Ensembles. The proposed strategy's privacy is guaranteed by limiting the teacher votes for student training, and by adding random noise to the topmost votes. The random noise is added under a well-established and rigorous standard of differential privacy, so that the privacy impact can be analyzed and bounded for each data item, and moments accountant technique is used to tighten the privacy bound. To further reduce the student/teacher consultation during the learning process, the authors presented how to do semi-supervised learning using generative adversarial networks (GANs). The experiment evaluation showed that the proposed scheme achieves an (ϵ, σ) differential privacy bound of $(2.04, 10^{-5})$ for MNIST with accuracy of 98%.

Ma et al. [57] introduced a different method to realize privacy-preserving multiparty learning. The proposed scheme (denoted as *SML* in [57]) also relies on knowledge transfer to construct labeled data from the aggregation of n locally trained model. Each participant trains a local model independently on the local dataset. Given a new public unlabeled dataset, those local models are aggregated together to generate label for each data item in the unlabeled dataset. But the aggregation is realized in a cryptographical manner. For each unlabeled date item, the participants give a classification (vote) based on their local models. A proxy re-encryption scheme is used to encrypt each participant's vote, which can later be summed up by a cloud server on the ciphertexts. Vote stands for the output of the local learning model upon the data item extracted from the auxiliary unlabeled dataset. *SML* achieves identical accuracy to the corresponding multiparty learning scheme proposed by Hamm et al. [40]. *SML* provides privacy guarantee for the local model and data, but these private information are all leaked to a trusted server in the scheme proposed in [40]. Moreover, the participants can verify the correctness of the returned result, thereby providing stronger security guarantee.

Bonawitz et al. [13] focused on mobile devices where communication is expensive and the workers dropout occur frequently and proposed two secure aggregation variant protocols which are low-communication overhead and robust to failures. The first one is secure against honest-but-curious adversaries but more efficient while the latter one can guarantee privacy against active adversaries with the cost of requiring an extra round. The main idea behind their design is double masking, i.e., based on the threshold secret sharing, two blinding factors are leveraged to conceal the original input data which is split and distributed to the remaining workers. The first blinding factors are generated via Diffie-Hellman protocol and form a circle naturally and then can be removed in aggregation phase. The second blinding factors are chosen randomly by each worker and required to reconstruct the secret by the remaining workers in the aggregation phase when workers revoke. However, the aggregate result is revealed to the server. Performance evaluation results show it is efficient, i.e., the communication cost and the running time remain low on large datasets and workers pools.

To reduce the high computation costs of those privacy-preserving machine learning schemes constructed from secure multi-party computation or fully homomorphic encryption, Ohrimenko et al. [64] proposed secure outsourced learning with trusted SGX processors in USENIX 2016. According to the introduction of SGX in [59], it is a set of new $\times 86$ instructions that applications can use to create enclaves, which are protected memory regions and isolated from any other code in the system. Only code running in an enclave can access data in the enclave. In the proposed system, the data owners agree on the machine learning tasks and on an SGX-enabled server to run the task. The machine learning algorithm is deployed into the processor-protected memory region. During the process, data owner uploads the encrypted data to non-protected memory but uploads their encryption keys into the SGX processor. The framework guarantees that only the machine learning code inside the trusted processor has direct access to the data, and the learned model will be returned to each data owner in encrypted form. The experiment results show that the computation costs of the proposed scheme are several orders of magnitude better than previous methods based on advanced secure multiparty computation or fully homomorphic encryption schemes.

Collaborative Learning (Multiple Non-colluding Servers) In [20], Chase et al. proposed private collaborative learning based on multiparty and differential privacy. The scheme was constructed on private gradient decent algorithm and collaborative gradient computation algorithm. In the collaborative setting, it is important to compute a differentially private gradient over the distributed records. If each data owner computes the local gradient from its local data and adds random noise to make it differentially private, the final learning model computed from the average gradients might have bad performance, since the amount of noise added will be large. In this scheme, the computed local gradients are secretly shared over two non-collusion servers H_1 and H_2 using secret sharing, who use secure two-party computation to output a differentially private gradient over the distributed records. The authors prove that the collaborative private gradient decent algorithm is differentially private so that the final learning model is also differentially private.

Mohassel and Zhang presented SecureML [62] that realize secure collaborative neural network learning under the two-server model. In the proposed protocol, the data are split into two additive shares and sent to two non-colluding servers, respectively. The two servers will complete the training interactively using secure two-party computation. New techniques were introduced in this paper, such as arithmetic on shared decimal numbers, multiparty computation friendly activation functions, and vectorizing the protocol in shared setting. For fix-point multiplication, decimal numbers are represented as shared integers in a finite field. Offline-generated multiplication triplets are used to perform a multiplication on shared integers. In addition, they proposed a new activation function which can be efficiently computed using garbled circuits. They use additively HE and OT to realize the multiplication on shared matrixes and vectors. The proposed protocol is secure against the honest-but-curious adversary who can corrupt a subset of clients but at most one of the two servers. Efficiency analysis showed that the protocol is highly efficient and total non-interactive since most of computations are offloaded to the two cloud servers.

In [61], Mohassel et al. presented a general framework (ABY) for outsourced deep learning in a three-server model. Training data are secretly shared among the three servers using secret sharing. And the learning task is offloaded to the servers who train the learning model through secure multiparty computation. During the three-party learning process, an efficient switching between arithmetic, binary, and Yao three-party computation will be used to accelerate the learning speed. In addition, new approximation technique for fixed-point multiplication over shared decimal numbers in three-party setting is proposed, which is much more efficient than previous methods and secures against malicious adversaries under the assumption that there is a single malicious adversary (server). The experiment evaluation shows that ABY is up to $55,000\times$ faster than SecureML when training neural network.

Most of the previous work on secure multiparty neural network learning focus on either applying secure techniques to existing training algorithms or on applying generic secure protocols to tailor training algorithms. These methods have several downsides, such as high offline computation overheads and online communication costs, accuracy loss and omitting the techniques necessary for modern deep learning. In [2], Agrawal et al. proposed QUOTIENT—a new method for multiparty learning in honest-but-curious two-server model. The scheme optimizes both the training algorithm and the secure protocol. The network weights are ternarized as $-1, 0, 1$ during the training process. Accordingly, a specialized scheme for ternary matrix-vector multiplication is constructed from correlated OT, Boolean sharing, and additive sharing. In addition, the operations of quantization and normalization in the backward process are replaced by alternatives which can be efficiently implemented in secure computation model without accuracy loss. Compared to SecureML [62], QUOTIENT obtains an improvement of $50\times$ in WAN time and 6% in accuracy (Table 7).

4.6 Privacy-Preserving Inference Outsourcing

Privacy-preserving inference outsourced is a new paradigm in deep learning outsourced. As shown in Fig. 5, three parties are involved in this new framework, including a model provider, the servers, and clients. In this new paradigm, the model

Table 7 Comparison of collaborative learning outsourced schemes with multiple servers

Publication	Main concept	Interactiveness	Security model
Chase et al. [20]	– Secretly share local gradients to two non-colluding servers. – Using 2PC to generate differentially private global gradients over distributed data. – Differentially private learning model.	Interactive	Honest-but-curious
Mohassel et al. [62]	– Share the training data to two non-colluding servers. – Arithmetic on shared decimal numbers. – Multiparty computation for activation functions. – Vectorizing.	Non-interactive	Honest-but-curious
Mohassel et al. [61]	– Privacy-preserving machine learning in a three-server model with a single corrupted server. – Efficient switching between arithmetic, binary and Yao three-party computation in the three-party setting. – Fixed-point multiplication for shared decimal numbers in three-party setting.	Non-interactive	Malicious
Agrawal et al. [2]	– Two non-colluding servers. – Optimize both the training algorithm and the secure protocol. – Ternarize network weight and ternarized matrix-vector multiplication in secret sharing setting.	Non-interactive	Honest-but-curious

provider offers a well learned and encrypted model to the server who can provide inference services to the clients. For privacy guarantee, the learned model and clients' data should be encrypted before uploading to the cloud. The cloud server provides inference services on encrypted model and encrypted data. Finally, the client will receive an encrypted classification which can be decrypted by the client or decrypted together by the client and the model provider. And when the prediction is done, the server should learn nothing about the private data and learning model. Meanwhile, the clients should learn nothing except the final prediction.

A quite similar research domain is oblivious neural network prediction [9, 18, 44, 51] in which the model owner himself/herself provides prediction services to the clients. And "oblivious" in this application scenario means that the predictions can be computed in such a way that the clients learn nothing about the model except the prediction results while the server knows nothing about the client's input data.

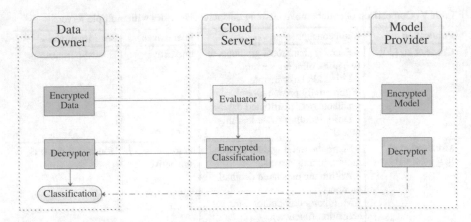

Fig. 5 The simplified framework of privacy-privacy inference outsourcing

In [56], Ma et al. presented a new secure outsourced prediction scheme that was secure against two non-colluding servers. The model provider securely outsources the pre-trained model to two non-colluding servers who can provide prediction service to data owners. To protect model privacy, the model owner splits it into two parts using additive secret sharing and uploads each part to one server, respectively. During the prediction phase, data are encrypted using additively homomorphic uploaded to both servers. Then, the servers evaluate the neural network on the ciphertext through secure two-party computation and return the encrypted classification back. The inference process is fully non-interactive for the data owner, and the proposed scheme is secure in honest-but-curious and non-colluding security model.

Jiang et al. [43] proposed a practical solution to secure outsourced matrix computation and apply it to privacy-preserving inference outsourced. Homomorphic encryption is used to encrypt the matrix and make it possible for an untrusted server to multiplication operations on encrypted matrices. For two matrices of size $d \times d$, the proposed matrix multiplication solution requires $O(d)$ homomorphic operations, compared with $O(d^2)$ of the previous works. The efficiency improvement is obtained from a new matrix encoding method and a combination of homomorphic operations and packed ciphertexts using the technique of Single Instruction Multiple Data (SIMD). The matrix computation mechanism can be applied in secure inference outsourcing assuming that the server is honest-but-curious. For the experiment evaluation, the authors choose the encrypted CNN model trained from MNIST, which includes one convolution layer, two fully connected layers with square activation functions. The experiment result shows that it requires 0.45 seconds per image, and the prediction accuracy reaches 98.1%. However, the framework works in the single-key scenario, which means that the network model and input data must be encrypted with the same encryption key.

Chen et al. [22] presented a new multi-key homomorphic encryption scheme. In the scheme, arithmetic operations on packed ciphertexts encrypted under different keys can be performed efficiently. A new relinearization key generation method,

Table 8 Comparison of privacy-preserving inference outsourced schemes

Publication	Main concept	Interactiveness	Security model
Ma et al. [56]	– Secretly share the pre-trained model to two non-colluding servers. – Use additive homomorphic to encrypt the data. – Use secure comparison to compute the activation function	Non-interactive	Honest-but-curious Non-colluding
Jiang et al. [43]	– Secure outsourced matrix multiplication. – Matrix encoding. – SIMD.	Non-interactive	Honest-but-curious
Chen et al. [22]	– A new multi-key homomorphic encryption. – Encrypts the pre-trained model and data with different keys.	Evaluation (non-interactive) Distributed decryption (interactive)	Honest-but-curious

which is more efficient and simplified compared to previous techniques, is proposed in this paper. The privacy-preserving inference outsourced scheme based on this multi-key homomorphic encryption is more efficient and flexible than previous schemes. The model provider encrypts the pre-trained model using his own public key and uploads it to a cloud server. Similarly, the data owner encrypts its data using his public key. The cloud server evaluates the neural network model on the ciphertext and outputs the encrypted classification. Then, the model provider and data owner work together to decrypt the ciphertext. Due to the multi-key homomorphic encryption, this inference service is suitable for the distributed application scenarios in which each data owner provides partial data to the server, and the prediction is evaluated on the aggregate data (Table 8).

5 Conclusion and Future Research Perspectives

In this paper, we bridge the gap between outsourced computation and deep learning, by investigating and presenting a comprehensive survey of the crossover between the two research areas. We briefly introduce the background of outsourced computation and deep learning. We then discuss the main techniques used in deep learning outsourced. Subsequently, we review and analyze the related works in outsourced deep learning. The related works are classified into individual training outsourced, collaborative training outsourcing, and privacy-preserving inference outsourcing.

As outsourced deep learning is achieving increasingly promising results, several important research issues remain to be addressed in the future. We conclude our survey by discussing these challenges.

1. *Collaborative Learning against Malicious Participant.* The main idea of federated learning is to build a machine learning model from multiple distributed datasets while preventing the raw data leakage. Typically, there is a parameter server who coordinates the distributed learning tasks, aggregates the response of the workers, and updates a global view of the model. Most of the previous researches on federated learning rely on the assumption that the data providers are honest. However, real-world applications often encounter such situations that some of the nodes might be unreliable or malicious, which means that some data providers might not behave as intended. Traditional federated learning becomes fragile when attacked by such Byzantine adversaries. Moreover, it has been proven that the parameter updates leak unintended information about participants' training data and develops passive and active inference attacks to exploit this leakage. Therefore, how to construct a federated learning system that is secure against Byzantine adversaries while preventing parameter updates leakage becomes an interesting problem.

2. *Secure Oblivious Neural Network Inference against Model Inversion Attack.* There are model inversion attacks against machine learning, including black box attack and white box attack. Black box attack requires only oracle access to a trained model, while white box refers to such attacks with full knowledge of the training mechanism and access to the model's parameters. In oblivious neural network inference schemes, black box attacks can be implemented by querying the server many times [76]. In [1], Abadi et al. proposed a deep learning scheme with differential privacy to tackle model inversion attack. Although differential privacy provides a strict privacy guarantee, it also brings accuracy loss. Therefore, how to guarantee privacy protection in oblivious neural network inference applications while preserving the prediction utility requires further research.

3. *Exploiting the Tradeoff among Effectiveness, Privacy, and Utility.* Privacy protection measures in outsourcing processes such as data encryption, blinding or perturbation might prevent the original data from being leaked at the cost of introducing utility loss and lower efficiency. The main impact on its efficiency influenced by the privacy protection mechanisms consists of computation overhead and communication overhead. Although homomorphic encryption and secure multiparty computation techniques provide the same meaning operations as in plaintexts, it inevitably makes great contributions to computation-intensive ciphertexts-based operations. Besides, compared to the traditional local training mode, secure outsourcing operations embedded in neural network training incur more interactive rounds. Besides, differential privacy involves relatively simple calculations. However, privacy protection is satisfied with sacrificing data utility. Hence, the next research direction is to seek the equilibrium point under three standards rules including effectiveness, privacy, utility. Accordingly, constructing

a relatively comprehensive privacy assessment mechanism facilitates a better awareness concerning the three factors, and further improves the performance of outsourcing computation applied to machine learning.

4. *Implementation of the Customized Privacy Protection for Clients.* Massive data contributes to a more accurate model, leading to the rise of federated learning or collaborative learning. What they have in common is that there are multiple data contributors or data providers. In these scenarios, only a fraction of users or some certain attributes need to be prevented from leakage. Therefore, a personalized privacy protection mechanism becomes a new research hotspot. Traditional outsourcing computation algorithms based on differential privacy, secure multiparty computation and homomorphic encryption techniques might not suitable in some application scenarios. Thus, we will devote ourselves to explore a dynamic perception of privacy requirements and the corresponding allocation of customized privacy budget in designing outsourcing algorithms. In this manner, the trained model can achieve high prediction accuracy while satisfying different privacy requirements for a variety of clients.

References

1. Abadi, M., Chu, A., Goodfellow, I., McMahan, H.B., Mironov, I., Talwar, K., Zhang, L.: Deep learning with differential privacy. In: Proceedings of the 2016 ACM SIGSAC Conference on Computer and Communications Security, pp. 308–318 (2016)
2. Agrawal, N., Shahin Shamsabadi, A., Kusner, M.J., Gascón, A.: Quotient: two-party secure neural network training and prediction. In: Proceedings of the 2019 ACM SIGSAC Conference on Computer and Communications Security, pp. 1231–1247 (2019)
3. Alpaydin, E.. Introduction to Machine Learning. MIT Press, Cambridge, MA (2014)
4. Aono, Y., Hayashi, T., Wang, L., Moriai, S.: Privacy-preserving deep learning via additively homomorphic encryption. IEEE Trans. Inf. Forensics Secur. **13**(5), 1333–1345 (2018)
5. Arulkumaran, K., Deisenroth, M.P., Brundage, M., Bharath, A.A.: Deep reinforcement learning: a brief survey. IEEE Signal Process. Mag. **34**(6), 26–38 (2017)
6. Avriel, M.: Nonlinear Programming: Analysis and Methods. Courier Corporation, North Chelmsford (2003)
7. Backes, M., Fiore, D., Reischuk, R.M.: Verifiable delegation of computation on outsourced data. In: Proceedings of the 2013 ACM SIGSAC Conference on Computer & Communications Security, pp. 863–874 (2013)
8. Barbosa, M., Farshim, P.: Delegatable homomorphic encryption with applications to secure outsourcing of computation. In: Cryptographers' Track at the RSA Conference, pp. 296–312. Springer, New York (2012)
9. Barni, M., Orlandi, C., Piva, A.: A privacy-preserving protocol for neural-network-based computation. In: Proceedings of the 8th workshop on Multimedia & Security, MM&Sec 2006, Geneva, September 26–27, 2006, pp. 146–151 (2006)
10. Bellare, M., Goldwasser, S., Lund, C., Russell, A.: Efficient probabilistically checkable proofs and applications to approximations. In: Proceedings of the Twenty-Fifth Annual ACM Symposium on Theory of Computing, pp. 294–304 (1993)
11. Benabbas, S., Gennaro, R., Vahlis, Y.: Verifiable delegation of computation over large datasets. In: Annual Cryptology Conference, pp. 111–131. Springer, New York (2011)
12. Blaze, M., Bleumer, G., Strauss, M.: Divertible protocols and atomic proxy cryptography. In: Proceedings of the International Conference on the Theory and Applications of Cryptographic Techniques, pp. 127–144. Springer, New York (1998)

13. Bonawitz, K., Ivanov, V., Kreuter, B., Marcedone, A., McMahan, H.B., Patel, S., Ramage, D., Segal, A., Seth, K.: Practical secure aggregation for privacy-preserving machine learning. In: Proceedings of the 2017 ACM SIGSAC Conference on Computer and Communications Security, pp. 1175–1191 (2017)
14. Boneh, D., Gentry, C., Lynn, B., Shacham, H.: Aggregate and verifiably encrypted signatures from bilinear maps. In: Advances in Cryptology - EUROCRYPT, pp. 416–432 (2003)
15. Boneh, D., Goh, E.-J., Nissim, K.: Evaluating 2-DNF formulas on ciphertexts. In: Theory of Cryptography Conference, pp. 325–341. Springer, New York (2005)
16. Boura, C., Gama, N., Georgieva, M., Jetchev, D.: CHIMERA: combining Ring-LWE-based fully homomorphic encryption schemes. Technical report, Cryptology ePrint Archive, Report 2018/758 (2018). https://eprint.iacr.org/2018/758
17. Brakerski, Z., Gentry, C., Vaikuntanathan, V.: (leveled) fully homomorphic encryption without bootstrapping. ACM Trans. Comput. Theor. 6(3), 1–36 (2014)
18. Brickell, J., Porter, D.E., Shmatikov, V., Witchel, E.: Privacy-preserving remote diagnostics. In: Proceedings of the 2007 ACM Conference on Computer and Communications Security, CCS 2007, Alexandria, VA, October 28–31, 2007, pp. 498–507 (2007)
19. Catalano, D., Fiore, D.: Practical homomorphic macs for arithmetic circuits. In: Annual International Conference on the Theory and Applications of Cryptographic Techniques, pp. 336–352. Springer, New York (2013)
20. Chase, M., Gilad-Bachrach, R., Laine, K., Lauter, K.E., Rindal, P.: Private collaborative neural network learning. IACR Cryptol. ePrint Archive 2017, 762 (2017)
21. Chen, X.: Introduction to secure outsourcing computation. Synth. Lect. Inf. Secur. Priv. Trust 8(2), 1–93 (2016)
22. Chen, H., Dai, W., Kim, M., Song, Y.: Efficient multi-key homomorphic encryption with packed ciphertexts with application to oblivious neural network inference. In: Proceedings of the 2019 ACM SIGSAC Conference on Computer and Communications Security, pp. 395–412 (2019)
23. Chillotti, I., Gama, N., Georgieva, M., Izabachène, M.: TFHE: fast fully homomorphic encryption over the torus. J. Cryptol. 33(1), 34–91 (2020)
24. Damgård, I., Geisler, M., Krøigaard, M.: Homomorphic encryption and secure comparison. IJACT 1(1), 22–31 (2008)
25. Deng, L.: A tutorial survey of architectures, algorithms, and applications for deep learning. In APSIPA Transactions on Signal and Information Processing, vol. 3 (2014)
26. Dwork, C.: Differential privacy: a survey of results. In: International Conference on Theory and Applications of Models of Computation, pp. 1–19. Springer, New York (2008)
27. Dwork, C., Kenthapadi, K., McSherry, F., Mironov, I., Naor, M.: Our data, ourselves: privacy via distributed noise generation. In: Annual International Conference on the Theory and Applications of Cryptographic Techniques, pp. 486–503. Springer, New York (2006)
28. Elgamal, T.: A public key cryptosystem and a signature scheme based on discrete logarithms. IEEE Trans. Inf. Theory 31(4), 469–472 (2003)
29. Fahlman, S.E.: Faster-learning variations on back-propagation: an empirical study. Proceedings of the Connectionist Models Summer School Morgan Kaufmann (1988)
30. Fredrikson, M., Lantz, E., Jha, S., Lin, S., Page, D., Ristenpart, T.: Privacy in pharmacogenetics: an end-to-end case study of personalized warfarin dosing. In: 23rd USENIX Security Symposium (USENIX Security 14), pp. 17–32 (2014)
31. Fredrikson, M., Jha, S., Ristenpart, T.: Model inversion attacks that exploit confidence information and basic countermeasures. In: Proceedings of the 22nd ACM SIGSAC Conference on Computer and Communications Security, pp. 1322–1333. ACM, New York (2015)
32. Gao, J., Fan, W., Jiang, J., Han, J.: Knowledge transfer via multiple model local structure mapping. In: Proceedings of the 14th ACM SIGKDD International Conference on Knowledge Discovery and Data Mining, pp. 283–291 (2008)
33. Gennaro, R., Wichs, D.: Fully homomorphic message authenticators. In: International Conference on the Theory and Application of Cryptology and Information Security, pp. 301–320. Springer, New York (2013)

34. Gennaro, R., Gentry, C., Parno, B.: Non-interactive verifiable computing: outsourcing computation to untrusted workers. In: Annual Cryptology Conference, pp. 465–482. Springer, New York (2010)
35. Gentry, C., Boneh, D.: A Fully Homomorphic Encryption Scheme, vol. 20. Stanford University, Stanford (2009)
36. Gilboa, N.: Two party RSA key generation. In: Advances in Cryptology - CRYPTO '99, 19th Annual International Cryptology Conference, Santa Barbara, CA, August 15–19, 1999, Proceedings, pp. 116–129 (1999)
37. Goodfellow, I., Bengio, Y., Courville, A.: Deep Learning. MIT Press, Cambridge, MA (2016)
38. Graves, A., Mohamed, A.-R., Hinton, G.: Speech recognition with deep recurrent neural networks. In: Proceedings of IEEE International Conference on Acoustics, Speech and Signal Processing, pp. 6645–6649 (2013)
39. Gu, S., Holly, E., Lillicrap, T., Levine, S.: Deep reinforcement learning for robotic manipulation with asynchronous off-policy updates. In: 2017 IEEE International Conference on Robotics and Automation (ICRA), pp. 3389–3396. IEEE, New York (2017)
40. Hamm, J., Cao, Y., Belkin, M.: Learning privately from multiparty data. In: Proceedings of the 33nd International Conference on Machine Learning, pp. 555–563 (2016)
41. Hao, M., Li, H., Xu, G., Liu, S., Yang, H.: Towards efficient and privacy-preserving federated deep learning. In: ICC 2019-2019 IEEE International Conference on Communications (ICC), pp. 1–6. IEEE, New York (2019)
42. Hinton, G., Deng, L., Yu, D., Dahl, G.E., Mohamed, A.-R., Jaitly, N., Senior, A., Vanhoucke, V., Nguyen, P., Sainath, T.N., et al.: Deep neural networks for acoustic modeling in speech recognition: the shared views of four research groups. IEEE Signal Process. Mag. **29**(6), 82–97 (2012)
43. Jiang, X., Kim, M., Lauter, K., Song, Y.: Secure outsourced matrix computation and application to neural networks. In: Proceedings of the 2018 ACM SIGSAC Conference on Computer and Communications Security, pp. 1209–1222 (2018)
44. Juvekar, C., Vaikuntanathan, V., Chandrakasan, A.: GAZELLE: a low latency framework for secure neural network inference. In: Enck, W., Felt, A.P. (eds.) 27th USENIX Security Symposium, USENIX Security 2018, Baltimore, MD, August 15–17, 2018, pp. 1651–1669. USENIX Association, Baltimore (2018)
45. Kilian, J.: Improved efficient arguments. In: Annual International Cryptology Conference, pp. 311–324. Springer, New York (1995)
46. Konečný, J., McMahan, H.B., Yu, F.X., Richtárik, P., Suresh, A.T., Bacon, D.: Federated Learning: Strategies for Improving Communication Efficiency (2016). Preprint. arXiv:1610.05492
47. Krizhevsky, A., Sutskever, I., Hinton, G.E.: ImageNet classification with deep convolutional neural networks. In: Advances in Neural Information Processing Systems, pp. 1097–1105 (2012)
48. LeCun, Y., Bengio, Y., Hinton, G.: Deep learning. Nature **521**(7553), 436–444 (2015)
49. Li, P., Li, J., Huang, Z., Li, T., Gao, C.-Z., Yiu, S.-M., Chen, K.: Multi-key privacy-preserving deep learning in cloud computing. Fut. Gener. Comput. Syst. **74**, 76–85 (2017)
50. Litjens, G., Kooi, T., Bejnordi, B.E., Setio, A.A.A., Ciompi, F., Ghafoorian, M., Van Der Laak, J.A., Van Ginneken, B., Sánchez, C.I.: A survey on deep learning in medical image analysis. Med. Image Anal. **42**, 60–88 (2017)
51. Liu, J., Juuti, M., Lu, Y., Asokan, N.: Oblivious neural network predictions via miniONN transformations. In: Proceedings of the 2017 ACM SIGSAC Conference on Computer and Communications Security, CCS 2017, Dallas, TX, October 30–November 03, 2017, pp. 619–631 (2017)
52. Liu, W., Wang, Z., Liu, X., Zeng, N., Liu, Y., Alsaadi, F.E.: A survey of deep neural network architectures and their applications. Neurocomputing **234**, 11–26 (2017)
53. López-Alt, A., Tromer, E., Vaikuntanathan, V.: On-the-fly multiparty computation on the cloud via multikey fully homomorphic encryption. In: Proceedings of the Forty-Fourth Annual ACM Symposium on Theory of Computing, pp. 1219–1234 (2012)

54. Lou, Q., Feng, B., Fox, G.C., Jiang, L.: Glyph: fast and accurately training deep neural networks on encrypted data (2019). Preprint. arXiv:1911.07101
55. Ma, X., Zhang, F., Chen, X., Shen, J.: Privacy preserving multi-party computation delegation for deep learning in cloud computing. Inf. Sci. **459**, 103–116 (2018)
56. Ma, X., Chen, X., Zhang, X.: Non-interactive privacy-preserving neural network prediction. Inf. Sci. **481**, 507–519 (2019)
57. Ma, X., Ji, C., Zhang, X., Wang, J., Li, J., Li, K.-C.: Secure multiparty learning from aggregation of locally trained models. In: International Conference on Machine Learning for Cyber Security, pp. 173–182. Springer, New York (2019)
58. Matsumoto, T., Kato, K., Imai, H.: Speeding up secret computations with insecure auxiliary devices. In: Conference on the Theory and Application of Cryptography, pp. 497–506. Springer, New York (1988)
59. McKeen, F., Alexandrovich, I., Berenzon, A., Rozas, C.V., Shafi, H., Shanbhogue, V., Savagaonkar, U.R.: Innovative instructions and software model for isolated execution. In: HASP@ ISCA, vol. 10(1) (2013)
60. Micali, S.: CS proofs. In: Proceedings 35th Annual Symposium on Foundations of Computer Science, pp. 436–453. IEEE, New York (1994)
61. Mohassel, P., Rindal, P.: ABY3: a mixed protocol framework for machine learning. In: Proceedings of the 2018 ACM SIGSAC Conference on Computer and Communications Security, pp. 35–52 (2018)
62. Mohassel, P., Zhang, Y.: SecureML: a system for scalable privacy-preserving machine learning. In: Proceedings of the 2017 38th IEEE Symposium on Security and Privacy (SP), pp. 19–38. IEEE, New York (2017)
63. Nandakumar, K., Ratha, N., Pankanti, S., Halevi, S.: Towards deep neural network training on encrypted data. In: Proceedings of the IEEE Conference on Computer Vision and Pattern Recognition Workshops (2019)
64. Ohrimenko, O., Schuster, F., Fournet, C., Mehta, A., Nowozin, S., Vaswani, K., Costa, M.: Oblivious multi-party machine learning on trusted processors. In: 25th USENIX Security Symposium (USENIX Security 16), pp. 619–636 (2016)
65. Paillier, P.: Public-key cryptosystems based on composite degree residuosity classes. In: Advances in Cryptology - EUROCRYPT '99, International Conference on the Theory and Application of Cryptographic Techniques, Prague, May 2–6, 1999, Proceeding, pp. 223–238 (1999)
66. Papernot, N., Abadi, M., Erlingsson, U., Goodfellow, I., Talwar, K.: Semi-supervised knowledge transfer for deep learning from private training data (2016). Preprint. arXiv:1610.05755
67. Parno, B., Raykova, M., Vaikuntanathan, V.: How to delegate and verify in public: verifiable computation from attribute-based encryption. In: Theory of Cryptography Conference, pp. 422–439. Springer, New York (2012)
68. Rumelhart, D.E., Hinton, G.E., Williams, R.J.: Learning internal representations by error propagation. Technical report, DTIC Document (1985)
69. Schmidhuber, J.: Deep learning in neural networks: an overview. Neural Netw. **61**, 85–117 (2015)
70. Shamir, A.: How to share a secret. Commun. ACM **22**(11), 612–613 (1979)
71. Shan, Z., Ren, K., Blanton, M., Wang, C.: Practical secure computation outsourcing: a survey. ACM Comput. Surv. **51**(2), 1–40 (2018)
72. Shokri, R., Shmatikov, V.: Privacy-preserving deep learning. In: Proceedings of the 22nd ACM SIGSAC Conference on Computer and Communications Security, pp. 1310–1321 (2015)
73. Silver, D., Huang, A., Maddison, C.J., Guez, A., Sifre, L., Van Den Driessche, G., Schrittwieser, J., Antonoglou, I., Panneershelvam, V., Lanctot, M., et al.: Mastering the game of go with deep neural networks and tree search. Nature **529**(7587), 484 (2016)
74. Song, W., Wang, B., Wang, Q., Shi, C., Lou, W., Peng, Z.: Publicly verifiable computation of polynomials over outsourced data with multiple sources. IEEE Trans. Inf. Forensics Secur. **12**(10), 2334–2347 (2017)

75. Taigman, Y., Yang, M., Ranzato, M., Wolf, L.: DeepFace: closing the gap to human-level performance in face verification. In: Proceedings of the IEEE Conference on Computer Vision and Pattern Recognition, pp. 1701–1708 (2014)
76. Tramèr, F., Zhang, F., Juels, A., Reiter, M.K., Ristenpart, T.: Stealing machine learning models via prediction APIs. In: 25th USENIX Security Symposium (USENIX Security 16), pp. 601–618 (2016)
77. Yao, A.C.-C.: How to generate and exchange secrets. In: 27th Annual Symposium on Foundations of Computer Science (SFCS 1986), pp. 162–167. IEEE, New York (1986)
78. Yu, L., Zhang, W., Wang, J., Yu, Y.: SeqGAN: sequence generative adversarial nets with policy gradient. In: Thirty-First AAAI Conference on Artificial Intelligence (2017)
79. Yu, X., Yan, Z., Vasilakos, A.V.: A survey of verifiable computation. Mob. Netw. Appl. **22**(3), 438–453 (2017)
80. Yuan, J., Yu, S.: Privacy preserving back-propagation neural network learning made practical with cloud computing. IEEE Trans. Parall. Distrib. Syst. **25**(1), 212–221 (2013)
81. Zhang, Y., Yang, Q.: A survey on multi-task learning (2017). Preprint. arXiv:1707.08114
82. Zhang, Q., Yang, L.T., Chen, Z.: Privacy preserving deep computation model on cloud for big data feature learning. IEEE Trans. Comput. **65**(5), 1351–1362 (2015)
83. Zhang, X., Jiang, T., Li, K.C., Castiglione, A., Chen, X.: New publicly verifiable computation for batch matrix multiplication. Information Sciences (2017). https://doi.org/10.1016/j.ins.2017.11.063

Printed in the United States
by Baker & Taylor Publisher Services